'Graph It !,
How to Make, Read,
and Interpret Graphs

Richard W. Bowen

Loyola University of Chicago

PRENTICE HALL, Englewood Cliffs, New Jersey 07632

Library of Congress Cataloging-in-Publication Data

Bowen, Richard.
    Graph It! : how to make, read, and interpret graphs   /   Richard W.
Bowen.
      p.   cm.
    Includes bibliographical references (p.   ) and index.
    ISBN 0-13-405515-2
    1. Graphic methods. I. Title.
QA90.B667 1992
001.4'1226—dc20

                           91-31948
                                CIP

For Nancy, David, and Laura

Acquisitions editor: Susan Finnemore Brennan
Production editor: Elaine Lynch
Copy editor: Donna Kuhn
Editorial assistant: Jennie Katsaros
Cover designer: Ray Lundgren Graphics, Ltd.
Cover graph concept: Richard W. Bowen
Interior graphs: Richard W. Bowen
Pre-press buyer: Kelly Behr
Manufacturing buyer: Mary Ann Gloriande

 © 1992 by Prentice-Hall, Inc.
A Simon & Schuster Company
Englewood Cliffs, New Jersey 07632

Printed in the United States of America
10 9 8 7 6 5 4 3 2 1

ISBN 0-13-405515-2

Prentice-Hall International (UK) Limited, *London*
Prentice-Hall of Australia Pty. Limited, *Sydney*
Prentice-Hall Canada Inc., *Toronto*
Prentice-Hall Hispanoamericana, S.A., *Mexico*
Prentice-Hall of India Private Limited, *New Delhi*
Prentice-Hall of Japan, Inc., *Tokyo*
Simon & Schuster Asia Pte. Ltd., *Singapore*
Editora Prentice-Hall do Brasil, Ltda., *Rio de Janeiro*

Contents

Preface

This book will teach you the skills of graph making, graph reading, and graph interpretation. *Graph It!* aims to help you develop these important skills rapidly and painlessly. The book is intended as a college course supplement to be read over a period of a week or so. I wrote *Graph It!* because my experiences in teaching Research Methods and Experimental Psychology convinced me that many bright and motivated students lack the skills that define **graph literacy**. The subject of graphs comes up in some college courses, but seldom in enough depth to insure that students really feel comfortable with reading and understanding graphs in textbooks and research papers or making graphs for their own papers and reports. *Graph It!* is intended to fill this gap in a college education.

Using the book is easy. Chapters 1 through 4 are a systematic guide to becoming graph literate. Reading all of the material in these chapters is strongly recommended. Chapter 5 is optional and covers a number of special topics concerning graphs, some of which you may be asked to read.

I hope you find *Graph It!* to be clear, interesting, and informative.

Acknowledgments

I would like to thank Susan Finnemore, Senior Psychology Editor at Prentice Hall, for her support and encouragement. My wife, Nancy S. Gerrie, not only cheered me along, but also cast a critical eye on my prose. Neil Agnew, York University; Dr. Terry D. Blumenthal, Wake Forest University; L.E. Bourne, Jr., University of Colorado; David Dooley, University of California; James Mazur, Southern Connecticut State University; F. J. McGuigan, U.S. International University; Steve A. Nida, Franklin University; Ron Nowaczyk, Clemson University; Lynn A. Olzak, University of California; and Janet T. Spence, University of Texas served as reviewers for the publisher and made a number of detailed and constructive suggestions for improving the manuscript. The majority

of the graphs appearing here are based on fictitious data. In some cases,
I have plotted factual data and I acknowledge, in either the text or the
figure caption, the actual source of such data. Finally, I thank Dr. Lee
Edlefsen and Trimetrix, Inc. for supplying the Axum software I used to
create the graphs contained here. Figures created with Axum were
modified for publication using Freelance Plus software (Lotus Develop-
ment Corp., 1989).

R. W. B.
Chicago, Illinois

1 | Introduction

The Focus of the Book

Graphicacy and Graphobia

The Main Purposes of Graphs

Purpose 1: The Time Series Graph

Purpose 2: The Correlational Graph

Purpose 3: The Graph of an Experiment

Graphs and Visual Communication

The Plan of the Book

Chapter Summary

This book is about how to make, read, and interpret graphs. In the course of your education, you have seen graphs, but you may be uncertain as to how to use and understand graphs. The problem is that no one ever really taught you how to read a graph and extract information from it. Courses about graphs are seldom taught in school. This is a shame because graphs are an extremely useful means of presenting data. The purpose of this book is to teach you about graphs: how to make them, read them, and interpret them.

When you make a graph, you turn numbers into a picture, a visual display of data. Figure 1-1 gives an example of a line graph based on some data that I made up. The data are fictitious average prices for chicken and for eggs at four-year intervals during the period from 1960 to 1988. At a glance, we notice the "uphill" slope of the data points. This direct relationship between the price of eggs and the price of chickens reflects the effects of inflation over the years. From 1960 to 1988 the prices for both chickens and eggs have gotten higher because inflation boosts the costs of these goods. But, in detail, the data show fluctuations in the markets for chickens and eggs. In recent years, for example, the price of eggs has dropped while the price of chicken has increased. This might be due to changing dietary concerns. People shun cholesterol-laden eggs (which cuts demand and lowers prices for eggs) while increasing their consumption of low-fat chicken (which boosts demand and increases the price).

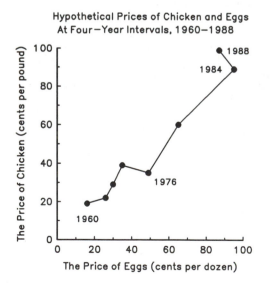

Figure 1-1 A graph of the relationship between the price of eggs and the price of chickens. Fictitious data values are given at four-year intervals from 1960 to 1988.

We could go on about this graph, discussing the factors that might influence the prices of chicken and eggs, such as general economic conditions (a recession, for example), natural disasters (such as a heat spell that killed off large numbers of domestic poultry), or public opinion (like the diet and health issue just mentioned). The point is that graphs are interesting, they help bring data to life. The visual nature of a graph helps us to remember data better and complements written descriptions of data sets. Looking at data in picture form is interesting and thought-provoking.

The effectiveness of a graph like Figure 1-1 is based upon the graph displaying numbers in a visual format. Reading a table of numbers (such as a list of chicken and egg prices) can be thought of as a **serial** process, in which the numbers must be checked **one at a time** to determine if they are getting larger, smaller, or staying the same. But a graph presents the numbers **in parallel**; all numbers are on display **at the same time**. It is always easier to read a graph than to read raw data in a table of numbers. It is also more interesting to read a graph than to merely read a verbal description of the data, such as "The price of chickens and the price of eggs tend to increase together, with some exceptions."

THE FOCUS OF THE BOOK

There are many different forms of graphs, including the line graph, the bar graph, the pie chart, and many other advanced special-purpose graphs (Cleveland, 1985; Tufte, 1983). In this book we concentrate on the simple line graph, such as the one in Figure 1-1. We focus on the line graph for several reasons. The line graph is the most common type of graph found in scientific, technical, and business fields. The line graph is the best all-purpose type of graph, since it can display the effects of several variables at once while remaining easy to read and interpret. Once line graphs are understood, other types of graphs become easier to understand. The line graph is the most elegant, but simple, visual technique known for presenting data.

GRAPHICACY AND GRAPHOBIA

Graphs are used extensively in fields of business and science. Graphs appear in high school and college textbooks, in technical manuals and scientific papers, in financial reports and on the business pages of the newspaper. In order to function in the modern technological world, a person must have a good level of **graphicacy** (graph literacy) (Balchin and Coleman, 1966; Schmid, 1983). Graphicacy is one of four critical skills that complete an education; the other skills are literacy (reading and writing), numeracy (math and numerical ability), and articulacy (communicating through speech). Graphicacy itself includes the skills of making, reading, and interpreting graphs that we intend to teach you in this book.

There are good reasons for students to become graph literate. One cannot understand science and technology in any real detail without referring to graphs. In scientific journals, the amount of space given to presenting graphs runs as high as 25 percent of the available pages (Cleveland, 1984). In business and finance, economic data are best understood through graphic presentation. The many fields of social science also rely on graphs to analyze research results.

For future progress, our educational system must provide a good basic background in science and technology. Access to scientific and technical knowledge demands graphic literacy. As it is, there appears to be a graphicacy gap between certain countries and the United States (Tufte, 1983). Leading Japanese newspapers, for example, print twenty times as many sophisticated graphs as *The New York Times* does.

Unfortunately, there is a high incidence of **graphobia** in this country among both students and workers. Many people have a "fear of graphs" because, as we have said, they have never really been taught how to read and interpret a graph. If you misread graphs in a textbook or business report, for example, or skip them altogether, you lose information, and you miss the point that the writer is trying to make. It is ironic that people shy away from graphs, since graphic representations of data are intended to be clear and simple. Graphs are intended to make it easy to read, understand, and remember a relationship found in a set of data. The purpose of this book is both to teach graphicacy and help cure graphobia.

THE MAIN PURPOSES OF GRAPHS

To get started, let's discuss how graphs are used. We have already noted that a graph is a means for converting numbers into a picture (a visual display or diagram). This is the most basic and important property of a graph. Numbers in a table must be examined one by one, but numbers represented on a graph can be seen all at once. Think of the line graph as an invention created to give a visual representation of an otherwise tedious list of numbers—the data. The human mind has great numerical ability, but it is not a good number checker. On the other hand, the human mind is extremely visual in nature, and graphs capitalize on the raw perceptual power that human beings possess.

Most line graphs are used for one of three purposes: to show a **time series** of data, to show a relationship or **correlation** between two variables, and to show a **causal connection** between variables that is discovered through an experiment.

Purpose 1: The Time Series Graph

The first "modern" graphs were time series graphs, and they were created over two hundred years ago by an English economist named William Playfair. In his book *The Commercial and Political Atlas* (1801), Playfair pioneered the use of graphs to display measures that change over time, for example, the price of wheat over a number of years, or yearly figures for the value of imports and exports to England (Tufte, 1983).

Playfair introduced the type of graph shown in Figure 1-2, which plots the population of the world since 1650. In such a graph, one can readily see trends developing in the measure. In this figure, we can easily see that the world population is not only growing, but growing at a steadily increasing rate. As we follow the data line relating population to time, we see the slope of the line increasing. We learn later on that the curve describing population growth is called an **accelerated relationship** as indicated by the fact that the slope of the curve (or rate of increase) is itself increasing. Beyond the simple reading of the graph, we might interpret this graph by focusing on the possible causes and consequences of accelerated population growth.

Plotting financial indicators or scientific data (such as interest rates or atmospheric ozone concentration) as a function of time is a common use

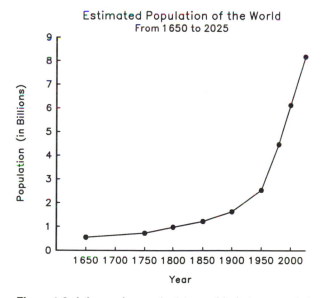

Figure 1-2 A time series graph of the world's human population over the past 100 years. Source of data: *The World Almanac and Book of Facts 1991.*

of graphs. Time series graphs help to anticipate the future, although the exact causes of trends over time may never be understood in great detail. The stock market is a perfect example of a time series of values that is so complex that people spend their lives charting the data and trying to predict future stock prices.

Purpose 2: The Correlational Graph

Much of the power of a two-dimensional line graph comes from its **comparative** or **relational function** (Tufte, 1983), the fact that it allows one to visually compare values of two or more variables. The term **variable** refers to any quantity or measure that can change and have different values (as we discuss in Chapter 2). A line graph naturally displays values of at least two variables at the same time, one variable represented by the horizontal (or X) axis, the other by the vertical (or Y) axis. Each data point on the graph represents two numbers or values (an X number and a Y number), each one coming from a different variable. (The chicken-egg graph in Figure 1-1 is a relational graph.) In a relational graph, the visual display of data points reveals how values of one variable change with changes in values of the other variable.

In Figure 1-3 we have a graph called a **scatterplot** that shows a relationship between two important personal variables: a person's height and weight. The graph plots values of the height and weight of each of

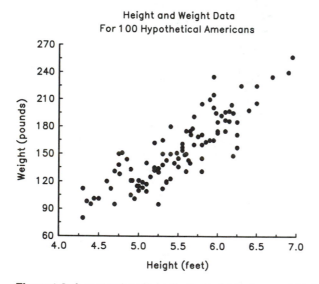

Figure 1-3 A scatterplot of hypothetical height versus weight data for 100 people. Each data point indicates both a given person's height (in feet, along the horizontal axis) and weight (in pounds, along the vertical axis).

100 hypothetical American citizens. Since each data point represents both the height and the weight of a person, this scatterplot displays a total of 200 separate numbers. If you were to read the individual height and weight values one pair at a time, it would be difficult to see any relationship in the data. When plotted on a graph, however, one can determine at a glance that the taller a person is, the heavier he or she is likely to be. We see this relationship clearly in the "uphill" slope of the cloud of data points. The relationship is clear even though various individuals are exceptions to the rule saying "the taller, the heavier" (tall people who weigh very little, short people who are heavy).

From Figure 1-3, we can say that height and weight are correlated. We use the term **correlation** to describe the fact that two variables change together, but may not directly influence one another. You might argue that getting taller causes a person to weigh more, but that connection is weak, since many tall people are light and many short people are heavy. For correlational data, whatever relationship is shown on the graph may be caused by variables other than those that are plotted. Height and weight are both influenced by other variables, including human genetics and individual dietary habits. When you simply measure two things at once (a correlation), there are many possible causes for the relationship on a graph of the data. As we noted before, the correlation between egg prices and chicken prices might be determined by a number of complex variables (the economy, the forces of nature, public opinion).

Purpose 3: The Graph of an Experiment

Line graphs are also used to display relationships discovered in experiments. An experiment is designed to discover whether there is a **cause-and-effect relationship** (a functional relationship) between two variables. Figure 1-4 is a graph of simulated data based on a classic type of experiment in my own field of experimental psychology, a reaction time experiment.

In a reaction time experiment, a human subject reacts to a stimulus (such as a flash of light) by making a response (such as pressing a button) as fast as possible. Meanwhile, the experimenter controls the brightness of each flash in steps from dim to bright and measures the reaction time to flashes at the different brightness levels. Reaction time is defined as the elapsed time from the start of the light flash until the subject presses the button. We all know that some people have "quick" reflexes (reaction times) while others are slow. But how will changing the brightness of the "go stimulus" affect the reaction time of any given person?

A look at the graph in Figure 1-4 tells us that reaction time gets shorter and shorter as the light flash gets brighter and brighter. The time to react decreases as flash brightness increases; we call this is an **inverse**

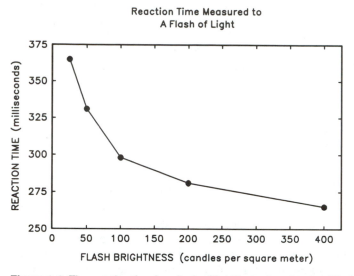

Figure 1-4 The reaction time to a flash of light for various levels of flash bright-ness. Reaction time is measured in milliseconds (thousandths of a second); brightness is measured by light units (candles per square meter).

relationship (as we learn in Chapter 3). We can also say that increasing flash brightness caused reaction time to decrease, since the relationship was discovered by experiment (see Chapter 4). What accounts for the relationship between the brightness of a stimulus and the time it takes to react to it? Perhaps brighter lights are processed faster by the nervous system than dimmer lights—reaction time is sometimes used as a mea-sure of the lag or latency of perception behind events in the real world (Bowen, 1981).

A correlation is a comparison, but an experiment is a **controlled** comparison. One variable is actively manipulated (we change the bright-ness of a flash) to discover the effect it has on a second variable (the reaction time). The resulting graph of data is a visual representation of the experimental effect. This is a sophisticated use for a graph, to chart a causal or functional connection between two variables.

GRAPHS AND VISUAL COMMUNICATION

For the past twenty years or so, people have been viewing the graph as an important means of visual communication. Edward Tufte (1983) and William Cleveland (1985) have pointed out that the effectiveness of a graph as a form of communication hinges on principles of graphical perception. The idea is that a graph is a complex visual stimulus, and human observers of the graph must extract the information available in the stimulus. Cleveland (1985) says that the graph maker encodes information, while the graph reader decodes it. If a graph is clear, uncluttered, and easy to read, communication between maker and reader is enhanced. If it is poorly designed or executed, communication will be reduced. If you think of a graph as an invention, a communication tool or technique, then you may appreciate that there are good tools and bad ones, and that a graph's overall performance as a visual display of data can always be improved by fine tuning.

THE PLAN OF THE BOOK

This book teaches you how to make, read, and interpret graphs. We have already defined what graphs are and what purposes they serve. In the next chapter we show you how to make a basic graph—a generic line graph that could be used for a class or could be included in a scientific, financial, or technical paper. The chapter lays out a step-by-step approach to creating a simple but sophisticated line graph, following principles of effective graphical design and perception.

In the third chapter, you learn to read graphs by recognizing basic relationships that can appear in a graph. The chapter covers direct versus inverse relationships, linear versus nonlinear relationships, monotonic versus nonmonotonic relationships, and other basic relationships that commonly appear in graphs. These concepts may sound difficult, but as you become familiar with them you will realize that you have encountered these relationships before.

In the fourth chapter we study the process of interpreting graphs, and arriving at an understanding of their meaning. Graph interpretation is based on recognizing the source of the data that appear in the graph, and the significance of the relationship depicted there. We discuss how to interpret time series graphs, correlational graphs, and graphs from experiments (the three applications of graphs introduced in this chapter).

The fifth and last chapter of the book covers special topics. We discuss special types of graphs, including frequency distributions, bar graphs, logarithmic graphs, and multiple-variable graphs. We also discuss statistical aspects of graphs. We conclude the chapter with an overview of computer graphics programs.

As you read along, bear in mind that this book is a primer on understanding graphs—the material presented here is basic stuff. Whether you learn it all very well now, or keep this book for quick reference later on, the book will help you in the goal of becoming graphically literate. By reading and comprehending *Graph It!*, you will graduate from simple time series graphs on the business page of the newspaper to the more abstract relational graphs that you find in textbooks and research papers. Let's turn now to Chapter 2, to learn how to create a basic line graph.

CHAPTER SUMMARY

This book is about how to make, read, and interpret graphs. Graphs are diagrams in which numbers are converted into a picture. The visual format of the graph makes it a superior tool for displaying, remembering, and analyzing data. There are three major purposes for graphs: to show time series data, to display correlations, and to present data from experiments. Access to scientific, technical, and financial fields requires graphicacy (graph literacy) although many people suffer needlessly from graphobia. In this book, you will learn how to construct a basic line graph (Chapter 2), how to read a graph by recognizing the relationship that the graph displays (Chapter 3), and how to interpret the significance of a graph (Chapter 4). You will also learn about special-purpose graphs, statistical aspects of graphs, and computer graphics programs (Chapter 5). Graphs have an advantage over tables of numbers in that they display numbers in parallel rather than as a series of entries. Graphs are powerful tools for visual communication, but communication can be either degraded or improved, depending on whether a graph is well designed or not.

2 | How to Make Graphs

PLAN OF THE CHAPTER

When you set out to make a line graph from a set of data, many basic questions may come up. How long do I make each axis? How should I divide each axis into a scale? How many "tick marks" should I use on the scale? What should I use for symbols when I plot the data? How should I label each axis? Or even, should I use a ruler?

The purpose of this chapter is to teach you how to make a graph. First you will see a graph with all of its parts labeled: a picture of the "anatomy" of a graph. You then learn ten simple steps to make a graph from a set of data, along with additional guidelines to improve the effectiveness of the graphs you make. Finally, there are some problems to test how well you have learned the material. You will get enough information from this chapter to produce a perfectly usable basic line graph, using only a pencil, graph paper and, yes, a ruler. Even though personal computers are taking over the work of creating graphs (Chapter 5, Section F), a graph-literate individual ought to be able to create graphs by hand as well as on the computer.

TEN EASY STEPS TO MAKING A LINE GRAPH

Figure 2-1 shows a line graph with its parts labeled. Look this graph over and read the labels for definitions of terms like **legend**, **data point**, and **axis scale**. We can make a graph like this in the following ten easy steps.

Step 1:	Identify the Data
Step 2:	Organize the Data
Step 3:	Find the Range for Each Variable
Step 4:	Draw the Axes
Step 5:	Design the Origin
Step 6:	Lay out the Scales
Step 7:	Put Numbers on the Scales and Label the Axes
Step 8:	Plot the Data Points
Step 9:	Add Lines to the Data Points
Step 10:	Add the Finishing Touches

Step 1: Identify the Data

Before putting pencil to paper, sit back and examine your set of data. You need to identify what kind of measures or values the data set contains.

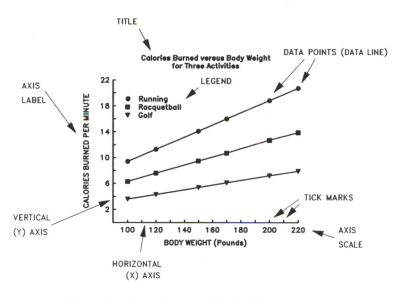

Figure 2-1 A sample graph with all of the parts labeled.

This will guide you in numbering and labeling axis scales later on and will help you to understand the graph that you create.

Table 2-1 gives examples of two kinds of data sets, quantitative (with numbers) and qualitative (with labels). Each data set contains values for two variables. A **variable** is defined as a quantity or measure that can assume different values. The height of a person, the temperature of the air, the population of a city, a state of the union, the color of a person's hair—these are all variables. If a variable is **quantitative**, its value is expressed by a number: height of 5 feet 11 inches, temperature of 68 degrees, population of 71,242 people. If a variable is **qualitative**, its value is expressed by a label: the state of "Ohio" or hair color that is "brown."

In a data set with two variables, values of the variables are given in pairs. In Data Set A of Table 2-1, each state (Variable 1) is paired with its particular value of average rainfall (Variable 2). In Data Set B, a person's height (Variable 1) is paired with his or her weight (Variable 2). To make a graph, one value in each pair is plotted along the horizontal axis and the other is plotted along the vertical axis. Each data point on the finished graph therefore stands for two pieces of information, a value of the horizontal-axis variable and a companion value of the vertical-axis variable. We symbolize each pair of values in a data set with the notation: (X, Y). The first value of the pair, X, is the value of the data point on the horizontal axis. The second value, Y, is the value on the vertical axis. If you have trouble remembering which axis is which, just think of the shape

Table 2-1

Data Set A

Variable 1 State of the Union	Variable 2 Average Annual Rainfall (inches)
Alabama	48
Arizona	12
Colorado	24
Louisiana	62
Nebraska	32
Wyoming	14

Data Set B

Variable 1 Height of a Person (feet)	Variable 2 Weight of a Person (pounds)
6.2	183
6.5	201
5.9	165
5.2	105
4.9	175
5.5	145

of the letter "Y" as being more vertical than the shape of the letter "X"—the bottom part of a Y points up and down, thus Y is the vertical axis. The Y axis is also called the "ordinate" while the X axis is called the "abscissa."

In a line graph, both the X and the Y variables are quantitative. In fact, a line graph is used to illustrate how some quantity along the Y axis varies as another quantity changes along the X axis. If you identify the X variable in your data set to be qualitative (like the states of the union, which are labeled with names and do not have numbers associated with them) you should stop! Don't make a line graph, instead make a bar graph (see Chapter 5, Section B). The reason is that if you "connect the dots" on a line graph by putting lines between data points, this implies that the horizontal scale was a scale of real amount (such as weight, height, the passage of time, or the intensity or strength of a stimulus). If the X variable is qualitative and doesn't indicate an amount, it is customary to represent the corresponding levels of the Y variable by the length of bars (the bar graph), rather than by the position of data points.   Note also that the Y variable is **always quantitative** in either a line graph or a bar graph because the purpose of a graph is to show a variation in the amount of the Y variable with a corresponding change (quantitative or qualitative) in the X variable.

Step 2: Organize the Data

Once you identify the quantitative or qualitative status of the X-axis variable, you should see how the data set is organized. It may be that the data you are given are not in any particular order. This won't stop you from plotting the data, as long as you have (X,Y) pairs of numbers. Nevertheless, you may want to rearrange the data to make plotting them easier and more accurate (in other words, Step 2 is optional but recommended).

Table 2-2

Disorganized Data		Organized Data	
X	Y	X	Y
10	17	2	7
6	9	4	10
2	7	6	9
8	12	8	12
4	10	10	17

The data set on the left of Table 2-2 is disorganized. To organize a data set, put the numbers for the X variable in numerical order from smallest to largest value, bringing along the Y value that belongs with each X value. The data set on the right of Table 2-2 has been organized in this way. We only put X values, not Y values, in order, because the graph will then show whether the quantity Y changes with some ordered change in the X variable.

Step 3: Find the Range for Each Variable

The next step is to determine the overall range that the X and Y variables fall into. The **range** of a variable is defined as the biggest number (MAX) minus the smallest number (MIN). The range equals MAX minus MIN.

Consider the data set in Table 2-3. Here the range for the X variable is 25 (120 minus 95) and the range for the Y variable is 60 (80 minus 20). The range of a variable will determine later on how to lay out the axis scales and how to number them. We will come back to the concept of the range in a little while when we talk about axis scales.

Table 2-3

X Variable	Y Variable
95	30
100	20
105	40
110	45
115	80
120	65

X range = 120 - 95 = 25

Y range = 80 - 20 = 60

Step 4: Draw the Axes

At this step, you finally take a sheet of graph paper and draw something: the horizontal and vertical axes, the framework of the graph. Graph paper provides a ready-made grid for the axes and number scales on the graph. Graph paper ruled with ten divisions to the one-half inch (10 X 10 to the 1/2 inch) is a good general-purpose graph paper, with enough divisions for accurate number plotting. You can purchase graph paper at any college book store and most stationery stores.

A simple rectangular design for drawing axes is shown in Figures 2-2 (a) and (b). A rectangular graph that is roughly 2 units high and 3 units wide has pleasing proportions and is easy to read. Stretching out the X axis gives us room to display up-down variations in the Y variable as the X variable increases. As shown in Figure 2-2, the 2-by-3 format can consist of just an X

The 2—by—3 Format

The Y Axis

The X Axis

Figure 2-2 Three general formats for laying out vertical and horizontal axes. In **(a)** the vertical axis is 2 units of distance, the horizontal axis is 3 (2-by-3 format).

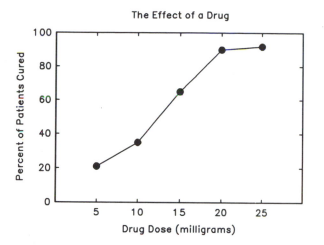

Figure 2-2 (b)   The vertical axis is 2 units of distance, the horizontal axis is 3 (2-by-3 format).   Shown is a data box with a sample completed graph.

axis at right angles to a Y axis, or we can draw a closed 2-by-3 rectangle to form a **data box**. Many authorities think that the data box is a superior design for a line graph (Schmid and Schmid, 1979; Cleveland, 1985); most computer graphics programs (see Chapter 5, Section F) give you a choice of using a data box or open X-Y axes. We use both formats in this book.

Figure 2-2 (c) shows the square format (2-by-2 format) for a graph. This is acceptable but seems a little less pleasing and harder to read.

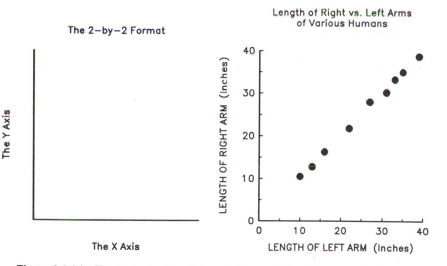

Figure 2-2 (c)   The square (or 2-by-2) format. If two variables change over the same range, the square format is usually a good design to use. In this format, you can easily tell whether the two variables change in exact proportion (the slope of the data line is 1.0) or not.

A square graph is a good choice if both the X and Y variables change over the same range. Most of the time in real graphs that won't happen; the two variables will change over entirely different ranges.

The format of 3 units high and 2 units wide (Figure 2-2 (c)) is an awkward proportion that seems to be "stretching" the vertical axis so that

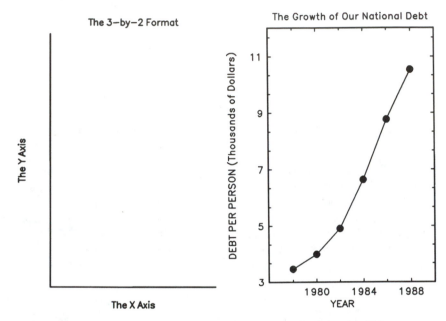

Figure 2-2 (d) The 3-unit vertical, 2-unit horizontal format (3-by 2-format). This format may tend to exaggerate the size (and significance) of changes in the variable on the vertical scale. As Tufte (1983) points out, William Playfair (the father of the modern graph) used the 3-by-2 format to chart the increase in the national debt of eighteenth-century England. The format is still exploited in that way.

vertical differences between data points will be larger. In fact, this format is commonly used to emphasize large changes in the Y variable for corresponding small or short-term changes in the X variable (Tufte, 1983), such as shocking increases in crime, taxes or the public debt over a short period of time.

Of the three formats, 2 units high and 3 units wide (or similar dimensions) will be the best choice for laying out the two axes. Most graph paper is 8 1/2 inches by 11 inches (standard letter size). If the horizontal axis follows the long side of the paper, a general size of 5 inches by 7 1/2inches (2 times 2 1/2 inches by 3 times 2 1/2 inches) for the axes will fit the paper with plenty of room to spare for axis labels, numbering, legends, and so on.

Step 5: Design the Origin

The **origin** of a graph can be defined as the point where the two axes intersect, in the lower left-hand corner of the graph. Before setting up the axis scales with tick marks and number values (in Steps 6 and 7) we need to decide what scale values the origin should represent. By looking at our data set and the variable ranges we already calculated, we must ask whether the origin is supposed to be a true zero-zero point, representing no amount of each of the two things being plotted. Alternatively, the lowest value to be represented on an axis may be something other than zero: a certain number of dollars or the earliest year in a series, for example.

In general, the origin will not need to be a zero-zero point. Many graph authorities believe that a zero point must be represented on an axis, even if it means putting a break in the axis by "cutting" it with a pair of short line segments, as shown in Figure 2-3 (a). For example, the American Psychological Association *Publication Manual* (1983) specifies, "If the graph could be misinterpreted because the origin is not zero, break the axes with a double slash" (p. 95). Follow that order if you must, but in my opinion, axis breaks are usually superfluous and always ugly. Graph readers who are paying attention do not have to be shown where zero is—they will automatically note how the origin of the graph is designed.

When we actually draw the origin at the axis intersection, we may want to set in the lowest value on each scale to the first tick mark on each axis, as shown in Figure 2-3 (b). The practice of setting in the minimum-value tick mark also prevents us from plotting data symbols right on the axis or at the origin, which could make the graph harder to read. Setting in the first tick mark away from the origin also makes it convenient to draw tick marks on the inside of the two axes, rather than on the outside. Tick marks on the inside look better and are better for reading off actual data values from the graph.

Step 6: Lay Out the Scales

At this step, we add the rest of the tick marks to the two axes to mark different values of each variable and create a scale. It is very important that each axis have a scale. Research on graphical perception (Cleveland, 1985) shows that the use of a common scale produces superior discrimination of the relative position of different data points on a graph.

The range of a scale on a graph should be somewhat greater than the range of the values of the variables involved. However, if the range of a scale is much greater than the range of the data to be plotted, this

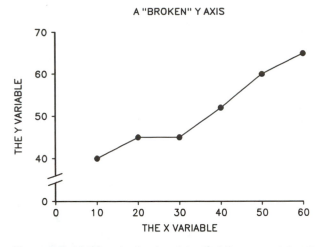

Figure 2-3 (a): Many textbooks advise that the zero point on the axis should be shown even if a scale doesn't start there. You're supposed to show the zero point and then put a "break" in the axis. My opinion is that the break is totally unnecessary: If you are paying attention to details while reading the graph, you will not need to be shown where zero is.

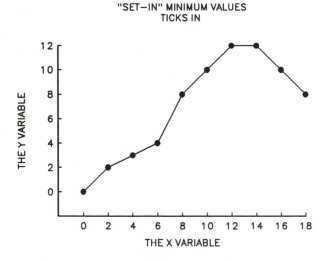

Figure 2-3 (b): If we set in the minimum value of each scale to the first tick mark, we can draw ticks on the inside of the axes, and we won't plot data symbols on an axis or at the origin.

will introduce visual distortion in the graph. Figure 2-4 illustrates good and bad ways to lay out vertical and horizontal scales. If there is a significant upward or downward trend in the data, using scales with ranges that correspond to the range of the data will plot data points along a line with a slope close to 2/3 (2 units up or down along the vertical axis versus 3 units across along the horizontal axis, assuming the 2-to-3 axis format we recommended previously in Step 4). Research on graphical perception (Cleveland, 1985) indicates that people are very poor at judg-

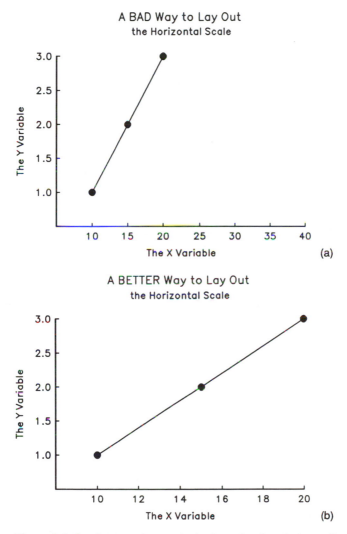

Figure 2-4 Good ways to lay out the horizontal and vertical axes (b, d) and bad ways to lay them out (a,c).

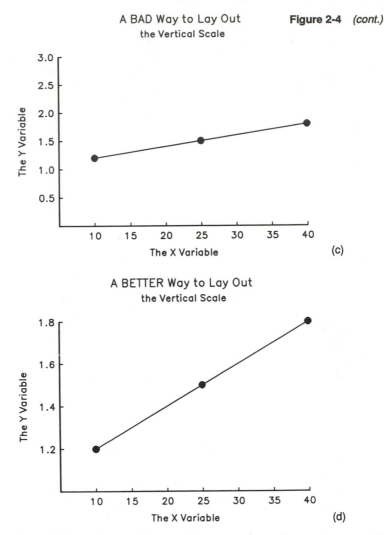

A BAD Way to Lay Out
the Vertical Scale

Figure 2-4 *(cont.)*

(c)

A BETTER Way to Lay Out
the Vertical Scale

(d)

ing the relative slope of lines with very steep slopes (near vertical) and very shallow slopes (near horizontal). The rules of Figure 2-4 keep the slope of data at middle values, not too shallow, not too steep, where discrimination of slope is very good (Cleveland, 1985).

Even though an axis needs to depict a scale of numbers, we do not want to show every possible value along that scale. Instead, we choose to divide each axis into manageable units using tick marks to create the scale of numbers.

One rule of thumb says that there should be at least three but no more than ten tick marks on each scale (Cleveland, 1985). For most graphs, about six tick marks per axis will be enough. Ticks should be located at easy-to-read intervals along the scale such as 5, 10, 50, and so on. On the other hand, if you want to plot a quantity that varies from year to year, and you are covering, for example, a fifteen-year range, you might want to let each year have its own tick mark.

Table 2-4

Flash Brightness	Reaction Time
(light units)	(milliseconds)
5	315
10	305
20	300
30	295
40	295
50	285

Table 2-4 contains a data set of measured reaction times to light flashes of various intensity levels, similar to the data plotted in Figure 1-4 of the first chapter. Let's take the data set in Table 2-4 and figure out how to lay out the two axes with tick marks and numbers. The X range in the example is 45 (50 minus 5). Using five tick marks, ticks pointing in, each pair of ticks should represent an interval of ten units. Furthermore, since the lowest X value is 5, we can let the X origin value be zero. The Y range is 30, minimum is 285, maximum is 315. The Y-axis origin value could be 270 in this case. On the Y axis, each pair of ticks should mark out 10 milliseconds. The two axes ought to be laid out in the same way as Figure 2-5.

The process of turning an axis into a particular number scale is somewhat arbitrary. For example, you want the person reading the scale to be able to estimate actual values of each variable from the location of the tick marks, so you need to include enough intervals to make that possible. But on the other hand, it is important to keep the scale as clear as possible, so too many intervals with too many tick marks are not good.

Step 7: Put Numbers on the Scales and Label the Axes

Don't forget to put numbers down next to tick marks so that the reader can tell what values are being represented. A number for every one or two tick marks is needed. Furthermore, each axis should be labeled with brief, easily understood descriptions of each variable. The units in which the variable is measured such as seconds, dollars, days, or inches, should also be clearly indicated in the label.

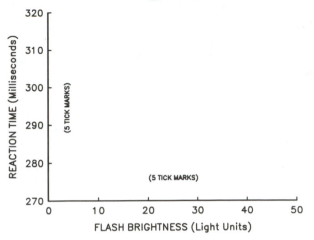

Figure 2-5 Laying out axis scales for the data of Table 2-4. With a range of reaction time from 285 to 315 milliseconds, we decide to start the scale at 270 milliseconds and extend it up to 320 milliseconds. The brightness scale, on the other hand, has a zero point, and stops at the highest light intensity used. The axis labels include the name of each variable and the units in which it is measured (units shown in parentheses).

When you draw graphs on tests or in reports, don't forget the all-important axis labels. Without labels on each axis, a reader can't tell what the graph is supposed to represent or describe. Look back to Figure 2-1 for examples of adequate axis labels.

Step 8: Plot the Data Points

At this step, we will actually plot the data set onto the graph. You will need to select a symbol to place at the (X, Y) coordinates of each data point. Usually the symbols are geometrical (circle, triangle, square, hexagon). Symbols can be made different sizes, and either "filled" or left "open."

Plotting the data points means finding the exact position that a data point should have in the graph. This also means that some kind of grid should be present to help plotting. I have already recommended graph paper with ten squares to the half inch as the best all-purpose graph paper.

Let's try plotting the reaction time versus flash intensity data from Table 2-4. Figure 2-5 has already shown us the appropriate layout of tick marks, numbers, and axis labels. We can use the "over right and up" rule for plotting data points. On the blank graph in Figure 2-6, the first data point from Table 2-4 is already plotted, point (5, 315) which is a reaction time of 315 milliseconds to a light of 5 units intensity. We see this "over right and up" maneuver indicated by the dashed arrows. The technique of first finding the correct X value (across) and then moving to the correct Y value (up) is something you usually learn in high school, although you may not have plotted any data points since then.

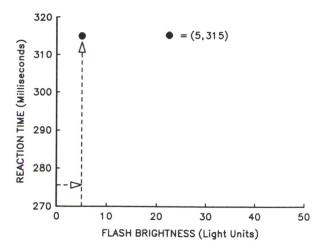

Figure 2-6 A set of axes for plotting the reaction time versus flash brightness data from Table 2-4. We have already plotted the data point (5, 315)—a reaction time of 315 milliseconds at a flash brightness of 5 units. Arrows indicate the "over right and up" maneuver for plotting. You should complete the graph from the table.

Let's proceed. Using Figure 2-6, finish plotting the data set. To plot a data pair, go over right to the X value on the horizontal axis, then up to the Y value on the vertical axis. Each data point you make can be a small circle about this size (o). When you have finished, compare your graph to the one in Figure 2-7. Are they the same? Good.

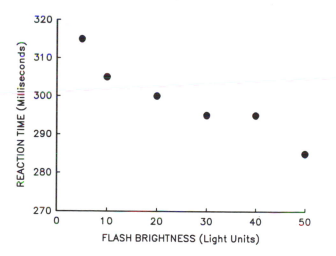

Figure 2-7 This is how Figure 2-6 should look after you have plotted the data from Table 2-4.

Step 9: Add Lines to the Data Points

The next step is to add lines in among the data points. The lines can simply connect the points, guiding the eye from one data point to another. This traces out how the values along the Y axis change with higher and higher X axis values.

Instead of connecting the data points, you could add a **regression line** (see Chapter 5, Section D). A regression line is a straight line representing the general position of all the data points, a "best fit" to the data. Figure 2-8 gives examples of both connecting the data points and adding a regression line.

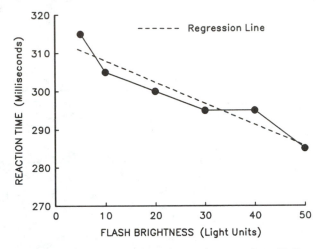

Figure 2-8 The same graph as the previous one, but with lines added to the data: simple connecting lines (the solid lines) or a regression line fit to the data.

Step 10: Add the Finishing Touches

The last things to add to the graph are additional text, title, legend, and so on to help make the graph easier to read. In Figure 2-9, we take the data from Figure 2-8, and add additional hypothetical data for a second subject in the experiment. Notice that we select a different data symbol for this additional relationship on the graph. Then we add the finishing touches. We put a title on the graph and a data point legend showing the plotting symbols are for each subject in the experiment. You should also look back one more time to the sample graph in Figure 2-1, in which we used a data point legend and a title. We're done! In ten steps, we have gone all the way from a table of numbers to a finished graph. This graph would be quite acceptable to submit with a laboratory report or research paper, and it even conforms to the style requirements of the

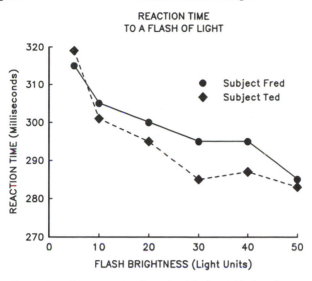

Figure 2-9 The same graph again, this time with data from another subject, a title, and a legend.

American Psychological Association *Publication Manual* (1983). Now we offer three simple guidelines to maximize the performance of the graphs you make.

THREE IMPORTANT GUIDELINES

Guideline Number One: Proofread the Graph!

Once you have made a graph, you should inspect it carefully before showing it to the public (or turning it in as a course assignment). A graph should be proofread as carefully as the text that the graph appears with. Indeed, it is essential that graphs be proofread to correct mistakes and to allow for improving the overall impact that a graph has. Even professionals sometimes fail to proofread graphs. It is estimated that 25 percent of published scientific research papers show graphs with mistakes and misprints in them (Cleveland, 1984; 1985). Most of the time, such mistakes in research papers could be caught and corrected with simple proofreading.

Guideline Number Two: Neatness Counts!

When you are preparing a graph for a report, paper, or course assignment, keep the graph as neat as possible. Use a ruler, erase extraneous marks, keep the size of numbers, labels, and legends in reasonable proportion to one another, and print all letters.

Guideline Number Three: Avoid Clutter!

A well-designed graph contains only essential information and min-imal decoration (Tufte, 1983). If a graph tries to portray too many variables at once, for example, or if the graph has an excessive number of labels or legends, it may become too cluttered to be effective and readable. Cleveland (1985) has stressed that graphs need "clear vision"—they must be uncluttered and designed to make the **data** the most prominent and visible components of the graph. Learning to make effective graphs is a skill that you develop with practice, as you create more and more graphs. You should now know enough about making graphs to start practicing on your own. You could begin by completing the following problem set.

PROBLEM SET

Now let's try plotting a few data sets. You can use the axes provided on the page, and you will need a ruler and a sharp pencil.

Problem 1

The following table shows data pairs on the performance of automo-biles, namely the size of a car's engine (in liters) versus the car's gas mileage (in miles per gallon). Plot the data on Figure 2-10 putting engine size along the X axis and gas mileage along the Y axis.

Car Make	Engine Size (liters)	Gas Mileage (miles per gallon)
Buick	3.2	21
Saab	2.1	30
Toyota	1.6	34
Cadillac	4.5	15
Volkswagen	2.0	23
Subaru	1.4	42

When you have plotted the data points, connect the points with straight lines. Notice that this graph can be read as showing a negative trend—gas mileage decreases as engine size increases. We can interpret this to mean that the bigger the car engine, the more gas required to go any given number of miles. A curve in which vertical-axis values decrease with increasing horizontal-axis values is described as an **inverse** or **negative** relationship, as we discover in the next chapter.

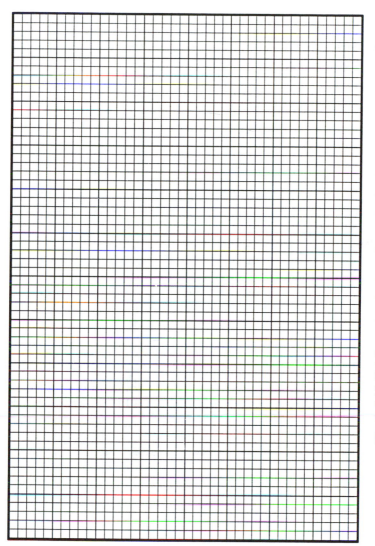

Figure 2-10 Plot the engine size-gas mileage data on these axes.

Problem 2

In the next table, there are hypothetical data for the hat size and the intelligence quotient of various human beings.

Hat Size	I.Q.
6	140
6 1/4	95
5 3/4	105
7 1/4	80
7 1/2	135
6 1/2	120
6 3/4	100

Plot the data points on Figure 2-11 and starting at the left of the horizontal (hat size) axis, connect successive points with straight lines, just as you did in Problem 1.

Your pencil sure does jump up and down, doesn't it? That's because there is no systematic trend in the relationship between hat size and I.Q. As you move up the horizontal axis in hat size, I.Q. jumps up and down, neither increasing or decreasing overall. In these hypothetical data, there is no effect—**no relationship** at all between the hat size a person wears and his or her measured intelligence quotient.

Problem 3

For the last problem, plot the data in the following table on Figure 2-12. The table has two variables: a person's height and shoe size.

Height (feet)	Shoe Size
5.5	8
6.0	11
5.9	11
6.5	12
7.1	14
5.0	7

As you can see from the plot, this is a **positive relationship**. Shorter people tend to have smaller shoe sizes, taller people tend to have larger shoe sizes.

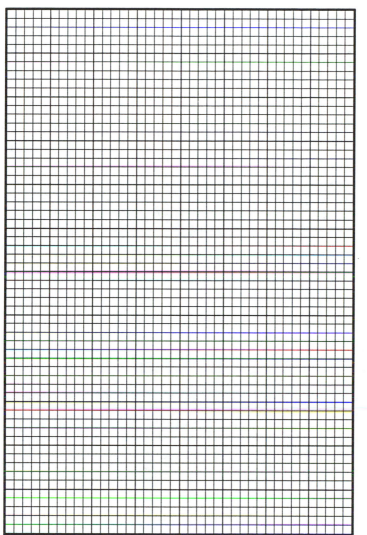

Figure 2-11 Plot the hat size-I.Q. data on these axes.

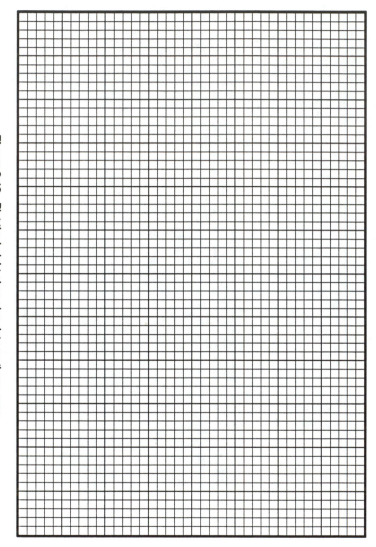

Figure 2-12 Plot the height-shoe size data on these axes.

CHAPTER SUMMARY

The chapter presents a ten-step procedure for making a graph from scratch:

Step 1:	Identify the Data
Step 2:	Organize the Data
Step 3:	Find the Range for Each Variable
Step 4:	Draw the Axes
Step 5:	Design the Origin
Step 6:	Lay Out the Scales
Step 7:	Put Numbers on the Scales and Label the Axes
Step 8:	Plot the Data Points
Step 9:	Add Lines to the Data Points
Step 10:	Add the Finishing Touches

Three general guidelines apply to graph construction:

Guideline 1:	Proofread!
Guideline 2:	Be neat!
Guideline 3:	No clutter!

PLAN OF THE CHAPTER

This chapter discusses the various classes of data relationships that may appear in a two-dimensional (X-Y) line graph, the kind of graph we learned to make in the previous chapter. The general goal of the chapter is for you to acquire some skill at reading line graphs. To read a graph is to identify, describe, and analyze the particular data relationships presented there. In this chapter you will learn to: (1) Identify or name the relationship presented in a line graph, based on its visual appearance, (2) describe the relationship by stating in words how the Y variable is changing with changes in the X variable, and (3) analyze the data relationship in terms of changes in the slope of the graph, where slope refers to how slanted a data relationship is, up or down, at any point on the graph.

Every graph of real data is unique, and the relationship that appears on the graph depends on the specific measurements that are plotted in the graph. In this chapter, however, we learn basic types of relationships, such as inverse, linear, and so on. Therefore, we show you a number of ideal graphs that consist of smooth curves or straight lines to illustrate the form of basic relationships. But real graphs, with real data, do not look so smooth. Real data points will deviate up and down, falling above or below a smooth curve representing an ideal relationship. In Chapter 5, we discuss the concept of **variability** in data. For the present, keep in mind the difference between real and idealized graphs.

THE GENERIC RELATIONSHIPS: Direct, Inverse, and Constant

Figure 3-1 shows the three most basic graphic relationships—direct, inverse, and constant. The first thing to notice is that the direct, inverse, and constant relationships differ in visual appearance. This visual difference boils down to the direction in which the data run on a graph. To identify these relationships, think of the direct relationship as going uphill, while the inverse relationship goes downhill. A constant relationship means no tilt; the data on the graph are level overall.

Second, each relationship is described in words differently. To describe the **direct relationship**, we say that **as the X variable increases, the Y variable increases**. The sample graph shows a direct relationship between years of schooling and average income. We read the graph by saying that the longer a person stays in school, the greater the income (on average) he or she can expect to make. To reiterate: as the X variable (years of school) increases, the Y variable (average income) also increases.

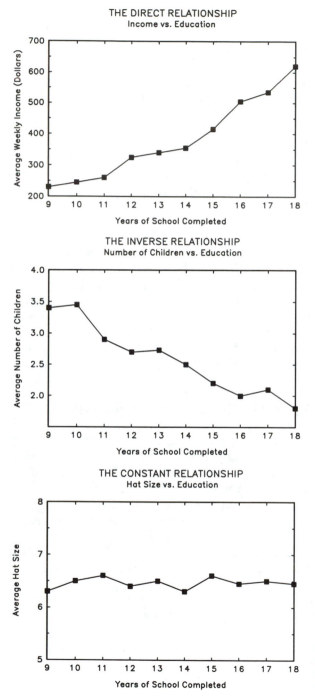

Figure 3-1 Graphs depicting the generic graphic relationships: direct, inverse, and constant.

We describe the **inverse relationship** by saying that **as the X variable increases, the Y variable decreases**. The sample graph for the inverse relationship can be read by saying that as a person's average number of years of schooling increases, the average number of children a person will have decreases. Here, as the X variable (the number of years of school) increases, the Y variable (the average number of children) decreases. We can also state that number of children in a family is inversely related to years of school completed by a parent.

For a **constant relationship**, we state that **as the X variable increases, the Y variable does not change**. A relationship will be called constant if overall there is no upward or downward trend in the data, even though the Y values are not all exactly the same. In the sample graph we show a relationship between years of school completed and average hat size. The data points form an essentially flat line: Overall there is neither an increase or a decrease in hat size as years of schooling increases. A constant relationship is essentially no relationship. In fact, we can even state that the value of the Y variable does not depend on the value of the X variable. In our sample graph, hat size (which is determined largely by genetic forces) is simply not affected by how much school a person completes.

The third aspect of reading these graphs is to analyze the slope of the data curve for each relationship. As we noted earlier, slope refers to the slant that a line has, up or down. In the direct relationship, the slope is positive because Y increases with X. In the inverse relationship, the slope of the graph is negative because Y decreases with X. And in the constant relationship, the slope is zero because there is no change in Y with X. As we will see later, the slope of a data curve can be given a numerical value, and can vary from shallow to steep. For the generic relationships, the direction of the slope is a fundamental property: These relationships can also be described as positive (direct), negative (inverse), or zero (constant).

THE LINEAR RELATIONSHIP

A quick look at a graph will tell us whether a relationship is direct or inverse, but we must take a somewhat harder look to decide whether a relationship is linear (straight-line) or whether it follows a nonlinear (curved-line) form. The linear relationship is the simplest form that data on a line graph can take; nonlinear relationships are more complicated to read and interpret.

Figure 3-2 shows two schematic linear relationships, one direct and one inverse. In terms of visual appearance, we identify these relationships as linear simply because in each graph the data points fall along a straight line. Note, however, that a straight data line means a linear relationship only if the scales of X and Y are arithmetic (spaced according to the ordinary number scale), as in Figure 3-2.

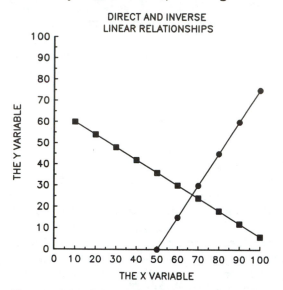

Figure 3-2 Idealized graphs of direct and inverse linear relationships.

Visual appearances can be deceiving—a data set that looks like a straight line may not be truly linear. A simple trick for telling by eye whether a data line is linear is to tilt the graph so that you can sight down the data line. This technique is illustrated in Figure 3-3, which shows that a graph that may seem to be linear in its normal orientation is clearly not a straight line when you look straight down the path of the data.

To verbally describe a linear relationship, we could simply say that the Y variable is linearly related to the X variable. A linear relationship implies that for any given change in the X variable, there is a constant change in the Y variable. This idea is illustrated in Figure 3-4 (a). No matter what value of X you start at, if you change X by some amount, Y will change by a fixed amount. This fact may seem obvious, but it is a powerful concept. Imagine, for example, that your grade on an exam was linearly related to the time you spent studying, as in Figure 3-4 (b). This would mean that a given increase in study time (say one hour) would result in a fixed increase in your exam grade (say 5 percent), a very desirable situation from a student's viewpoint. Relationships that are linear in form are simple and well-behaved relationships.

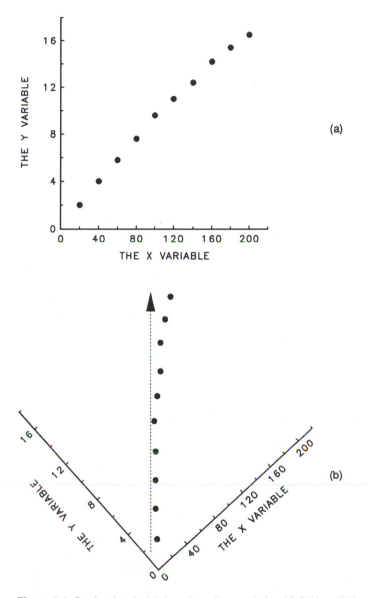

Figure 3-3 Do the data in **(a)** describe a linear relationship? If we tilt the graph and sight along the data line, as in graph **(b)**, we clearly see the points veering away from a straight-line relationship.

In a linear relationship, the fact that any change in X produces a constant change in Y leads us to define the most important aspect of a linear relationship: the slope of the straight line on the graph. The slope is determined by how much the Y variable changes for a given change in

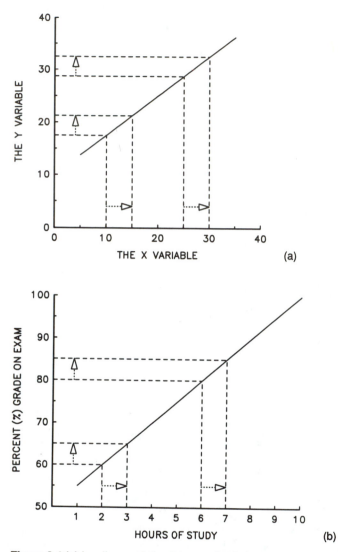

Figure 3-4 (a) In a linear relationship, any fixed change in the X variable results in a fixed change in the Y variable. **(b)** In this linear relationship between time spent studying and grade performance, any one-hour increase in study time produces a 5 percent grade increase.

the X variable. If you think back to high school math, you will recall that the slope of a straight line refers to how much a line "rises" (or "falls") relative to how much it "runs." To be precise, the numerical value of the slope of a line equals the amount of change in the Y variable for a given change in the X variable.

In Figure 3-5, we show several schematic linear relationships and we calculate the slope of the lines. In Figure 3-5 (a), the value of X changes by a standard 10 units, and the Y variable changes by 20 units. The slope is thus 20/10 which equals 2. In Figure 3-5 (b), the Y variable changes by 5 units for a 10-unit change in the X variable. The slope is 5/10 or 0.5. Note in Figure 3-5 (c) that if a relationship is inverse, the Y variable decreases some amount for a given increase in the X variable. And if the

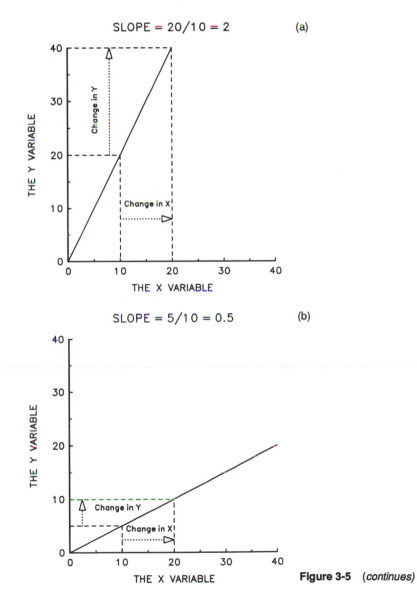

Figure 3-5 (continues)

SLOPE = −15/10 = −1.5   (c)

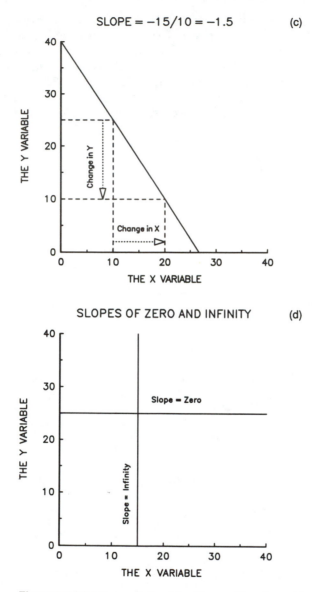

SLOPES OF ZERO AND INFINITY   (d)

Figure 3-5 (a) A linear relationship with a positive slope of 2.
(b) A linear relationship with a positive slope of 0.5. **(c)** A linear relationship with
a negative slope of 1.5. **(d)** Linear relationships with slopes of zero and infinity.

value of the Y variable falls rather than rises, then the data line will have
a negative slope. In Figure 3-5 (d) note also that if a line is flat, with no
change in Y for a change in X, then the slope is 0 (zero). And if a line on
a graph is completely vertical, the slope is defined as infinity, since for

essentially no change in the X variable, Y increases by a very large (infinite) amount.   We should also point out the simple fact that if a relationship is linear, the slope of the line is constant and does not change as the X variable increases.   In the majority of real graphs, the X and Y scales are not the same. They represent totally different variables, even though the two scales may have arithmetically spaced values. In that case, the slope of a data line is best expressed not as a single number, but as a fraction showing the measured change in the Y variable for a given change in the X variable.   For example, in Figure 3-6, we show a graph of a person's salary (in thousands of dollars) versus the number of years the person has worked at a job.   On this linear graph, we can readily determine that the salary has increased $10,000 in each 5-year period, a slope of $2,000/year.   Expressing the slope as dollars divided by years reflects the fact that two different variables are depicted on the graph.

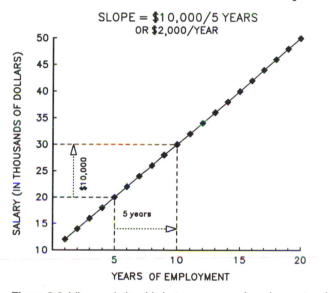

Figure 3-6 A linear relationship between years of employment and salary. The slope is $2,000/year.

NONLINEAR RELATIONSHIPS

Direct or inverse relationships can also be nonlinear in form. We can define a relationship as nonlinear by the fact that the slope of a data line on the graph is not constant (as in a linear graph) but is changing as the X variable increases. In Figure 3-7 (a) we show an idealized graph that is called a **direct accelerated** relationship. In this type of graph, the slope changes, starting out at a low value, and increasing as the X variable increases.

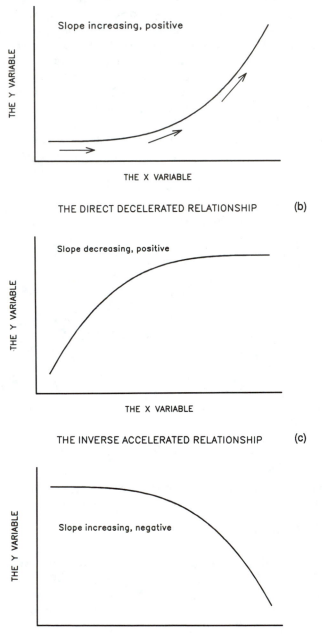

THE DIRECT ACCELERATED RELATIONSHIP   (a)

Slope increasing, positive

THE Y VARIABLE

THE X VARIABLE

THE DIRECT DECELERATED RELATIONSHIP   (b)

Slope decreasing, positive

THE Y VARIABLE

THE X VARIABLE

THE INVERSE ACCELERATED RELATIONSHIP   (c)

Slope increasing, negative

THE Y VARIABLE

THE X VARIABLE

Figure 3-7   *(continues)*

THE INVERSE DECELERATED RELATIONSHIP   (d)

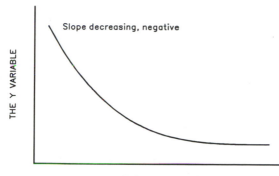

Figure 3-7 Four nonlinear relationships. **(a)** Direct accelerated. The relationship is direct (Y increases with increasing X) and accelerated (the slope increases with increasing X). **(b)** Direct decelerated.**(c)** Inverse accelerated.**(d)** Inverse decelerated.

As indicated on the figure, this change in slope can be visualized by drawing in a series of arrows tangent to the curved lines at various points. A tangent line indicates the slope at any one point on a curve. As we have said, the slope of a nonlinear relationship is constantly changing. The tangent arrows indicate that this direct accelerated relationship is changing from a slope of 0 (flat) to a steep slope. The term **accelerated** tells us that the slope is increasing, that a given change in X produces a bigger and bigger change in Y as the value of X increases. The term **direct** indicates that all tangent arrows on the curve are pointing up, the slopes are positive.

Figure 3-7 also shows idealized graphs of three other basic kinds of nonlinear relationships: panel (b) **direct decelerated**, panel (c) **inverse accelerated**, and panel (d) **inverse decelerated**. These differ simply in terms of whether the slopes are positive (direct) or negative (inverse) and whether the slopes are increasing (accelerated) or decreasing (decelerated). You should draw tangent arrows on the panels of the figure to indicate how slopes are changing. With your arrows, Figure 3-7 completely summarizes the visual appearance, verbal description, and slope changes for each type of nonlinear graph.

One way to think about these types of nonlinear relationships is that the slope change is in one direction only. That is, either the slope increases in absolute value over the graph (accelerated) or it steadily decreases in absolute value (decelerated). In decelerated relationships, the slope may eventually reach zero, a flat line. In that case, where the value of the Y variable stops changing, we say that the relationship has reached an **asymptote**.

The relationships shown in Figure 3-7 are sometimes called **monotonic** relationships. Monotonic means that the value of Y on the vertical axis always either increases steadily or decreases steadily with increases in the X variable along the horizontal axis. Later on we discuss **nonmonotonic** relationships, where the slope of the relationship changes direction (goes both up and down) as well as changing in steepness.

Now let's consider some examples of these nonlinear relationships.

Direct Accelerated and Direct Decelerated

For examples of direct nonlinear relationships, we turn to two different fields of psychology: social psychology and perceptual psychology.

Our social psychology example comes from an analysis of social impact by Bibb Latané (1981). Social impact refers to changes in attitude or behavior that are brought about by the presence of other people. Latané's theory suggests that the strength of social impact increases as the number of people present increases. However, each additional person contributes less and less to the social impact that a group has. The first person to be present has a greater impact than the tenth person.

Among many examples in support of these ideas, Latané cites an interesting vintage study of stuttering (Porter, 1939). In this experiment, persons identified as stutterers were asked to read a 500-word passage in front of audiences of different sizes (0, 1, 2, 4 or 8 members). The results showed that the greater the size of the audience, the greater the percent of words stuttered by the subjects, and thus the greater the social impact of the audience. Figure 3-8 shows the general form that this relationship followed. It is a direct decelerated curve. Consistent with Latané's theory, as persons are added to the audience, each has less and less impact, as indicated by a smaller and smaller increase in the percent of words stuttered by the persons experiencing the social impact of the audience. In a direct decelerated relationship, as the X variable increases, the Y variable also increases, but by progressively smaller amounts.

Our example from perceptual psychology is a nonlinear relationship in the perception of the strength or intensity of electric shock discovered by S. S. Stevens (1962). Any person perceives increasing discomfort or pain as an applied electric shock level is increased. The issue is exactly how the perceived intensity of a stimulus (a psychological variable) was related to the actual intensity (a physical variable).

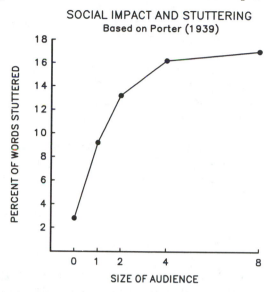

Figure 3-8 The relationship between the percent of words stuttered and the number of persons in an audience watching the stutterer. Based on Porter (1939) as analyzed by Latané (1981). An example of a direct decelerated relationship.

To measure a variable such as electric shock level, Stevens could simply use a physical instrument such as a meter. To measure perceived shock, he used human instruments or subjects to report on their perceptions. Specifically, he instructed human subjects to report how intense a shock seemed to be by assigning a number or magnitude estimate to the perceived discomfort. In perceptual psychology, this is called a **scaling** task; Stevens's version of scaling is also called **magnitude estimation**.

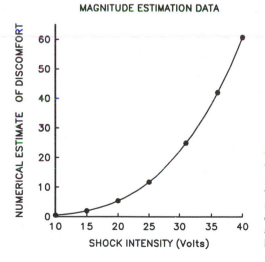

Figure 3-9 Magnitude estimates of the discomfort of electric shock plotted as a function of shock intensity. Perceived shock increases with intensity according to an accelerated relationship—discomfort or pain increases more and more for higher and higher levels of shock intensity. Based on Stevens (1962).

Figure 3-9 plots a magnitude estimation curve, based on numerical magnitude estimates of perceived shock intensity made by observers at various physical shock levels. The relationship is a direct accelerated one. Low-intensity shock increases perceived pain slowly, with a shallow slope. As shock level grows, perceived pain grows even faster, with an ever increasing slope to the function. In an accelerated relationship, each small change in the X variable produces larger and larger changes in the Y variable.

It is easy to see the survival value of the accelerated relationship of Figure 3-9. At higher shock intensity levels, larger and larger increases in perceived pain result from small changes in shock level. This fact helps a person to react forcefully if a noxious stimulus such as shock actually threatens to destroy tissue.

Inverse Accelerated and Inverse Decelerated

To take a fictitious but plausible example of inverse nonlinear relationships, suppose that you have a friend who is determined to lose some weight and signs up with a weight-loss program. The program guarantees that the friend will lose 18 pounds, going from 150 pounds to 132 pounds. The question is, how will the weight loss proceed over time?

Figure 3-10 shows three possibilities for a graph of your friend's weight (in pounds) versus weeks in the weight-loss program. The top curve indicates that the weight loss is at first gradual (shallow negative slope), but more and more pounds are lost as weeks go by (increasing

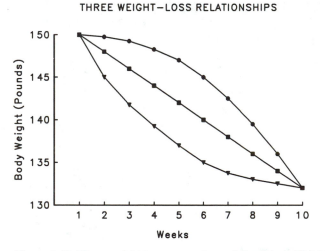

THREE WEIGHT–LOSS RELATIONSHIPS

Figure 3-10 Three weight-loss versus time relationships. Weight lost accelerates with weeks of dieting (open circles), weight lost is linear with weeks of dieting (squares), or weight lost is decelerated with weeks of dieting (diamonds).

slope, an inverse accelerated relationship). The middle curve is a linear relationship: Your friend loses the same number of pounds each week (constant negative slope). The bottom curve is an inverse decelerated relationship where the most weight is lost in the first week (steep slope), with gradually less and less weight lost in succeeding weeks (decreasing slope, a decelerated relationship).

A person in a weight-loss program may wish that pounds shed might follow a linear relationship, or even an accelerated curve, but the truth is that most people lose weight according to an inverse decelerated relationship. Weight loss is large at first, more gradual later on. If you have ever dieted, you know what its like on the long, lonely right-hand slope of the weight-loss functions, where it seems to take forever to lose those last few pounds.

In the social impact theory discussed previously, Latané (1981) analyzed an interesting case of an inverse decelerated relationship: the relationship between the size of a tip left by a group of diners and the number of diners in the group (Freeman, Walker, Borden and Latané, 1975). Real-life data on tipping were collected by waiters in a Columbus, Ohio, restaurant, and involved hundreds of groups of diners. The results are represented by the curve in Figure 3-11. As the number of diners at a table increases from one to six, the percent tip decreases from more than 18 percent to just over 12 percent. The relationship is decelerated. The size of the tip drops sharply from one diner to a twosome, but each additional diner produces smaller and smaller reductions in the size of the tip, consistent with Latané's theory.

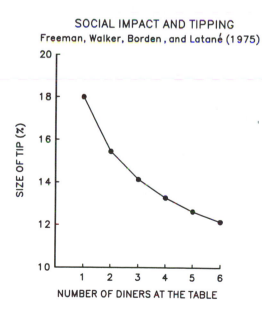

Figure 3-11 Another example of an inverse decelerated relationship. The size of a tip left at a restaurant meal (in percent) as a function of the size of the group of diners. Based on Freeman, Walker, Borden, and Latané (1975) as analyzed by Latané (1981).

THE S-SHAPED RELATIONSHIP

In the nonlinear relationships studied so far, the slope changes, but always in one direction, steep to shallow or shallow to steep. Another type of direct or inverse nonlinear relationship is the **S-shaped relationship**. Here the slope starts out shallow, gets steeper, then returns to being shallow. In visual appearance, the curve is shaped like the letter S, as the name implies. In the field of psychology, the most famous S-shaped relationship is known as the **psychometric function**. A psychometric function comes from a perceptual experiment in which subjects try to detect a stimulus (a light, for example) which is variable in intensity, from weak (not detectable) to strong (clearly detectable). A series of light levels is presented to the subject over and over again, and we record for each stimulus whether the subject said "Yes" (I see it) or "No" (I don't see a light). We then construct a graph from the data, as shown in Figure 3-12. This is a relationship between percent "Yes" reports (percent seeing) and the intensity of the light stimulus to be detected.

THE PSYCHOMETRIC FUNCTION

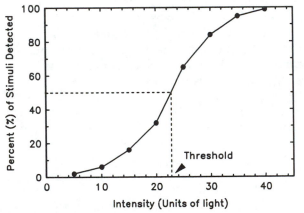

Figure 3-12 The psychometric function. The percent of light stimuli detected is an S-shaped function of the stimulus intensity. The slope of the relationship is shallow for weak lights or for strong ones, and the slope is steeper at intermediate intensity values. The **threshold for light detection** is defined as the intensity detected 50 percent of the time, as indicated by the dashed lines in the graph.

The psychometric function can be described verbally simply by calling it S-shaped. In a slope analysis of the S-shaped relationship, we first note that the slope at either end of the relationship is shallower than the slope in the middle of the graph. For a range of weak stimuli (low intensity), the percent of stimuli detected remains near zero (shallow slope): The subject is certain of not having seen a light. Similarly, for a range of strong stimuli (high intensity), the slope is also shallow and the subject is very certain (100 percent detection) of having seen a light.

In between weak and strong stimuli, the percent detection increases. It is interesting that the slope of the increase is not infinite (vertical line). Instead, there is a range of stimuli for which the subject is very uncertain. For example, from Figure 3-12, we see that a light of 20 units is detected about 30 percent of the time, while a light of 30 units is detected about 85 percent of the time. Why does the subject not say "No" to all lights of 20 units and "Yes" to all lights of 30 units?

One possible reason for the variability in the subject's detection responses is that from moment to moment the internal state of the subject's nervous system is changing. This internal noise produces **false positive** responses (the light was weak, but the response was "Yes") and **false negative** responses (the light was strong, but the response was "No"). We have all experienced this kind of uncertainty near **threshold** (the transition from not perceived to perceived). We have all asked questions like "Was that the phone ringing or not? Was that the baby crying or not?" The uncertainty evident in the S-shaped psychometric function has led psychologists to define the **threshold stimulus** as that intensity that is detected 50 percent of the time, as indicated by the dashed lines on Figure 3-12.

MONOTONIC AND NONMONOTONIC RELATIONSHIPS

Thus far all of the relationships we have learned to read in graphs have been monotonic in form. A **monotonic** curve is one in which the value of the Y variable steadily increases or steadily decreases with increases in the X variable. However, we may often come across nonmonotonic relationships in real data sets. In a **nonmonotonic** relationship, the Y variable can change in both up and down directions (the slope changes direction as well as steepness).

Figure 3-13 (a) shows several schematic graphs that are monotonic; Figure 3-13 (b) shows schematic graphs that are nonmonotonic. In non-monotonic data sets, Y values may change both up and down with increases in the X variable, causing dips and bumps in the data curve. As you might be able to guess from looking at Figure 3-12 (b), there is an infinite number of possible forms for nonmonotonic curves. On the other hand, we frequently come across the following two nonmonotonic relationships: the **U-shaped** and **inverted U-shaped** relationships.

As an example of the U-shaped relationship, we examine the **serial position curve**. In research on memory, it has long been known that when learning a long list of unrelated words, a person has better recall for words that come either first in the list or last in the list. Words in the

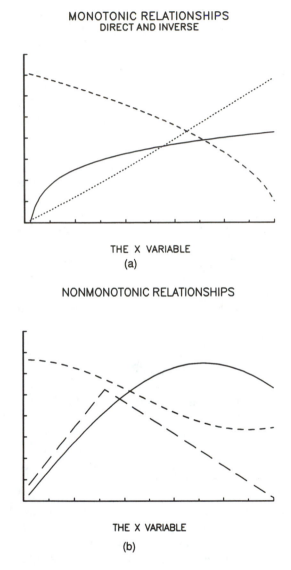

MONOTONIC RELATIONSHIPS
DIRECT AND INVERSE

THE Y VARIABLE

THE X VARIABLE
(a)

NONMONOTONIC RELATIONSHIPS

THE Y VARIABLE

THE X VARIABLE
(b)

Figure 3-13 **(a)** Some monotonic relationships. The slope changes in only one direction. **(b)** Some non-monotonic relationships. Slope values are both positive and negative over the range of the relationship.

middle of the list are recalled worst of all. This relationship is depicted in Figure 3-14 in a made-up graph of the percent of subjects who gave correct responses to the various items in the list. The relationship between level of recall and position of an item in a list is U-shaped (shaped like the letter U).

The two steep sides of the U are attributed to different processes in memory. The fact that words coming first in the list are recalled very well

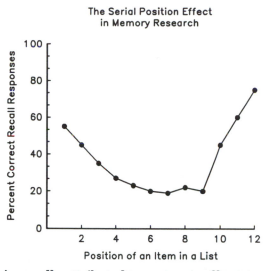

The Serial Position Effect
in Memory Research

Figure 3-14 The serial position effect. Percent of correct recall responses versus the serial position of an item in a list is a U-shaped relationship.

is usually attributed to processes of long-term memory, while superior recall at the end of the list is attributed to short-term memory (Baddeley, 1990).

As an example of the inverted U-shaped relationship, consider the well-known psychological relationship between level of performance and level of arousal or motivation. Suppose that before a big exam, I survey every student in the class to determine how motivated they are to do well on the exam. When the exam is over, I plot the self-confessed motivation level of each student versus the grade he or she received on the exam, as shown in Figure 3-15.

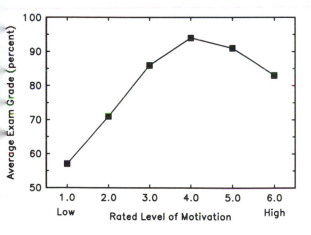

Exam Performance versus Motivation

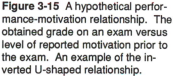

Figure 3-15 A hypothetical performance-motivation relationship. The obtained grade on an exam versus level of reported motivation prior to the exam. An example of the inverted U-shaped relationship.

The relationship between exam grade (performance) and motivation level is shaped like an inverted U. Increasing motivation from a low level produces steadily improving performance. Performance reaches a peak, however, and the exam grade falls if students are too highly motivated. Maybe you have found yourself on the down side of the inverted U: so "up" for an exam that your motivation level actually interferes with performance and your grade ends up lower than it could be.

THE BI-LIMBED RELATIONSHIP

In a bi-limbed relationship, there is not a single smooth curve, but two separate functions that intersect each other. Consider the following made-up example of a bi-limbed relationship. Figure 3-16 shows a graph of the size of a balloon (the diameter in inches) as a function of the number of puffs of air that have been blown into it. The relationship is bi-limbed, with a straight line of slope = 0 (no change in size up to four puffs of air blown in) and a direct decelerated curve (for five to twelve puffs of air). The two limbs imply that two processes are at work here. Up to four puffs, the pressure inside the balloon is not enough to overcome the resistance of the rubber walls of the balloon. For more than five puffs, the resistance is overcome and the balloon steadily inflates. We see the bi-limbed relationship in the two obviously separate data lines that describe the balloon's behavior, and we distinguish the lines based on their differing form.

A BI−LIMBED RELATIONSHIP

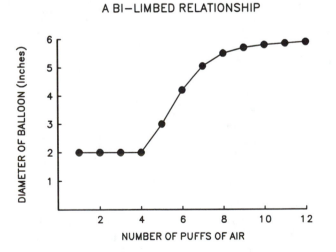

Figure 3-16 A hypothetical example of a bi-limbed relationship. The diameter of a balloon is graphed in relation to the number of puffs of air used to inflate it.

TIME SERIES RELATIONSHIPS

A **time series** is a graph of a **sequence of measures** of any real-world variable (such as temperature or stock market closing values). In a time series, we expect that values of the Y variable will fluctuate up and down with repeated measurements.

In a fluctuating graph, we take note of two visual features. First, we try to see if the local fluctuations in the graph are **periodic** (repeating over and over) or **random** (completely irregular in sequence). In a plot of stock market prices over time, it is sometimes argued that day-to-day fluctuations in value trace out a perfectly random sequence. On the other hand, daily temperature readings show a definite cycle, going down at night and coming up during the day.

The second thing to look for in a fluctuating time series is whether there is any overall trend upward or downward across the course of the graph. Here we ignore the local fluctuations, and look for a gradual rise or fall in the average or overall value of the Y variable. As an example, Figure 3-17 graphs real data on crime rates for the period from 1985 to 1989 (Source of the data: *The World Almanac and Book of Facts 1991*). In spite of some up-down fluctuation, it is clear from the graph that there has been an increase in violent crime (murder, rape, and so on) and a decrease in burglary over the five-year period.

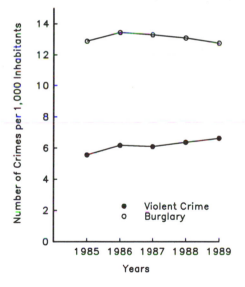

Figure 3-17 Trends in crime rate. Number of crimes per thousand U. S. inhabitants for 1985 to 1989. Violent crimes shown by filled circles, burglaries by open circles. (Source of the data: *The World Almanac and Book of Facts 1991*.)

In general, time series graphs are extremely useful for visualizing progressive change in any variable. In Chapter 4, we devote a section to the interpretation of time series relationships.

FAMILIES OF CURVES

When you read the relationship shown on a graph, you can also be thinking that the graph at hand belongs to a family of similar lines or curves. In imagining a family of curves, you are imagining possibilities, graphs that might have been plotted instead of the one that was.

As an example, consider the family of nonlinear relationships shown in Figure 3-18. These direct accelerated relationships represent the salary growth curves for three different people, all of whom were hired by a particular company at the same time, at a salary of $10,000. Each person receives yearly salary increases, but for each person the yearly percent is different (3 percent, 6 percent, or 9 percent) over the ten-year period in question. It is striking that after ten years Mary, who got 9 percent raises, is making almost $10,000 more than Chet, who got only 3 percent raises. This illustrates the power of compound interest, as the salary gap between Mary and Chet widens with each passing year.

A graph like Figure 3-18 illustrates the comparative function of line graphs. Not only does a line graph allow us to visualize the relationship

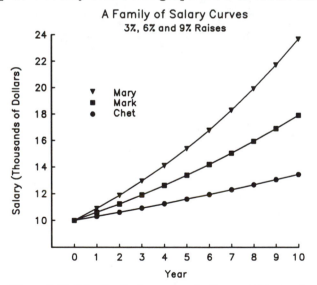

Figure 3-18 A family of salary curves for three employees who each start a job at $10,000 per year. All curves are direct accelerated relationships, representing 3 percent raises (for Chet, circles), 6 percent raises (for Mark, squares) or 9 percent raises (for Mary, triangles).

between two variables, but it allows us to compare two or more relationships on the same graph. (As a further example, look back at Figure 3-17 for a comparison of the recent history of the incidence of two types of crime.) We can invoke the comparative function of graphs whenever we view any single graph as a member of a family of possible relationships.

THE EFFECTS OF AXIS DISTORTIONS

As we have learned, the slope of a relationship is a critical feature which guides us in reading the relationship. Slope is also an indication of the **magnitude** or **size of an effect** displayed on a graph. A steep slope implies a big effect of X on Y, a shallow slope implies a small one.

But the slope that we visually assess on a graph depends on how the X and Y scales are laid out, and what overall range of numbers are represented on each axis. The same data set can produce different visual assessments, depending on how the two axes of a graph are constructed. In Figure 3-19, we show how the apparent size of a difference between two data points can be made to change, by either stretching the Y axis or shrinking the X axis. Either manipulation increases the slope of the relationship, and gives the impression that the effect plotted is a larger one. These manipulations are possible violations of **graphical integrity** (Tufte, 1983) since they may mislead the reader, giving the impression that a small effect in data is a significant one.

In general, when you are either making graphs or reading them, it is important to pay attention to the scales of both axes. When you inspect a relationship with a steep slope, you must decide whether the change is a significant one. The best way to determine the significance of a slope or a difference in data points is with a separate statistical analysis of the data. If there are no statistical indicators (such as **error bars**—see

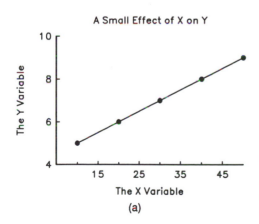

A Small Effect of X on Y

The Y Variable

The X Variable

(a)

Figure 3-19 Examples of distortions in graphs. A small effect (shallow slope) is made to look bigger (steeper slope) by either stretching the vertical axis of the graph (b) or shrinking the horizontal axis of the graph (c).

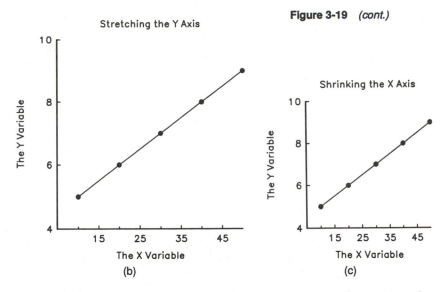

Figure 3-19 *(cont.)*

Chapter 5, Section D), you must take care to examine the axis scales themselves to judge whether a steep slope appearing on the graph is in reality not a large effect. This completes our survey of graphic relationships. Before going on to Chapter 4, you should try the following quiz.

A GRAPH READING QUIZ

To find out how well you learned to read graphs, take the following quiz.

     1. Calculate the slopes of the linear relationships shown in Figure 3-20.

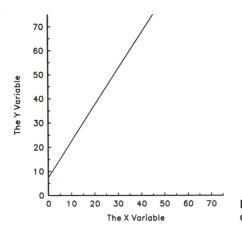

Figure 3-20 Calculate the slopes of these linear relationships.

(**Figure 3-20** *(cont.)*

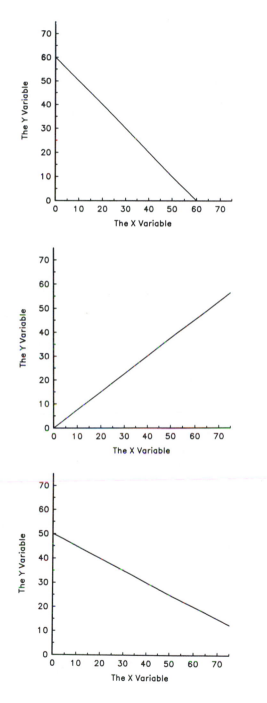

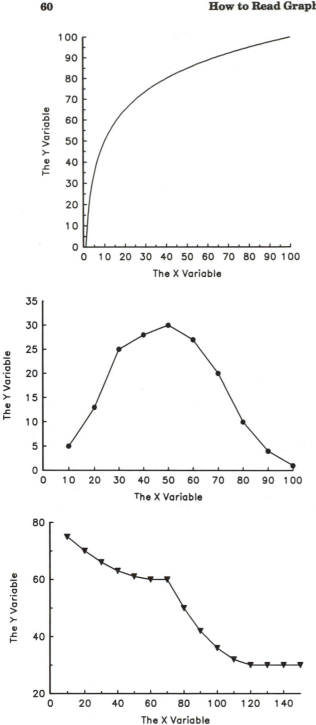

Figure 3-21 For these graphs, name the relationship, describe the relationship, and state how the slope of the relationship changes.

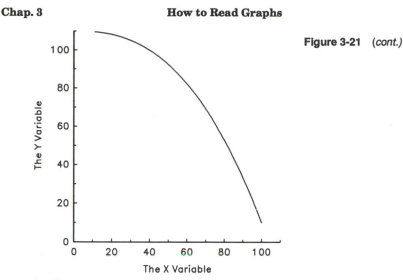

Figure 3-21 *(cont.)*

2. For graphs shown in Figure 3-21, **name** the relationship shown, **describe** the relationship, and state how the **slope** of the relationship changes.

CHAPTER SUMMARY

To read a graph is to identify the relationship displayed based on visual appearance, describe how the Y variable depends on the X variable, and analyze the changes in slope in the graph. The generic relationships are called direct (the data line has a positive slope), inverse (the graph has a negative slope), or constant (the graph has a zero slope). The linear (straight-line) relationship has a constant slope, but in nonlinear relationships the slope changes. There are four simple nonlinear relationships: direct accelerated, direct decelerated, inverse accelerated, and inverse decelerated. They differ in the direction of the curve and whether the slope is increasing or decreasing as the X variable changes. In an S-shaped relationship, the slope changes from shallow to steep and back to shallow. Nonmonotonic relationships are defined by changes in slope that go both up and down. In the U-shaped relationship, the slope of the graph is first steep and negative, then goes to zero, then becomes steep and positive. In the inverted-U graph, the slope changes are reversed. In a bi-limbed relationship, there are two separate relationships present, differing in form, in slope, or in position on the graph. In a time series relationship, we focus on whether local variation in the Y variable is periodic or not and on whether there is an overall trend up or down in the graph. Any single graph may belong to a family of graphs and the graph at hand is only one of many possible outcomes. Beware of distortions of the X or Y axis that can lead you to conclude that a steep slope in a relationship represents a significant effect when it actually does not.

How to Interpret Graphs

Plan of the Chapter

The Difference between Reading and Interpreting a Graph

Interpreting a Graph from an Experiment

Interpreting a Correlational Graph

Interpreting a Time Series Graph

Practices in Graphic Interpretation

Case 1 (Psychology): Practice, Practice, Practice

Case 2 (Sociology): Knowledge and Nuclear Power

Case 3 (Education): The Scholastic Aptitude Test—
Males, Females, and Math

Chapter Summary

PLAN OF THE CHAPTER

In Chapter 2, we learned how to make a graph. In Chapter 3, we covered reading graphs and the various forms of graphic relationships. In this chapter, we come to the final step, **graph interpretation**, understanding the meaning or significance of a graph. The remaining chapter in the book (Chapter 5) covers some special topics. While you may be asked to read some of these, they aren't part of the central focus of this book: making, reading, and interpreting graphs.

We begin this chapter by distinguishing between reading a graph (recognizing the relationship present in the data) and interpreting a graph (discovering the significance and implications of the relationship). The key to graph interpretation is understanding the source of the data that appear in the graph. We discuss the interpretation of graphs of data from three primary sources. These are (1) data derived from an **experiment** (examples: the results of a medical experiment; measurements of the performance of a piece of electronic equipment), (2) data derived from a **correlation**, where two variables are measured and compared (examples: radon levels versus lung cancer deaths; population density versus crime rate) and (3) data derived from a **time series**, where measures are taken at time intervals (examples: daily stock averages; world population over years). In the last section of the chapter, we practice what we have learned by interpreting sample graphs.

THE DIFFERENCE BETWEEN READING
AND INTERPRETING A GRAPH

Let us review briefly how reading and interpreting are defined. As we have learned, to read a graph is to state the relationship that is present, to describe the relationship, and to analyze changes in the slope of the data curve. To interpret a graph is to determine the significance of the data relationship that is present, including the causes of the relationship and the practical implications of it.

Consider this example of the difference between reading and interpreting a graph. Suppose that an astronomer has discovered a gigantic comet that is moving in the direction of the earth. She is able to estimate its distance from the earth each day for a period of two weeks. She plots the distance from the earth as a function of days, as shown in Figure 4-1. We read Figure 4-1 as showing an inverse linear relationship. The distance of the comet from the earth is inversely related to days of elapsed time and is decreasing linearly, by the same amount every day.

As you interpret the graph, on the other hand, you might well scream "WE ARE DOOMED!!!" This recognizes that as days go by into the future,

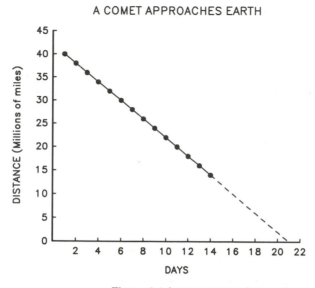

Figure 4-1 A comet approaches earth.

the distance between earth and the comet will grow even smaller, until finally there is a collision. Since the inverse relation is linear, we can easily predict the day of the future collision by extending the line from the present day. On Figure 4-1, this projection is shown as a dashed line which intersects the horizontal axis seven days from today. This prediction is part of the interpretation of the graph. So is the observation that the comet must be traveling at a constant speed, since the distance covered each day is the same.

As you can tell from this example, in interpreting a graph you expand upon the implications of the data. By contrast, reading a graph is usually a **closed-end** process. Once you identify a particular relationship, there is usually not much more to say about the form of the graph. Interpretation is **open-ended**; since once a discussion of the significance of a graph begins, it need never end. Interpretation of the data is an ongoing process that is always subject to reevaluation and restatement.

In graphic interpretation, we ask questions like "What is the message this graph is trying to convey? What factors are responsible for the relationship shown on the graph? What are the implications of this relationship?" When you ask such questions, you are looking for answers to satisfy your own logic and common sense. Furthermore, you should always take the intellectual offensive when you read and interpret graphs and act like a skeptic, an individual who needs to be convinced that the apparent message of a graph is real. The skeptical person will look for alternative interpretations of the relationship that is depicted in a graph.

The best and most basic "common sense" question to ask of a graph is "How were the data on the graph collected?" Having a clear understanding of the source of the measurements plotted on a graph is the best foundation for interpretation. If we understand the source of the data, we begin to get at the meaning of the data. Let us now consider the general interpretation of graphs based on three major sources of data: the experiment, the correlation, and the time series.

INTERPRETING A GRAPH FROM AN EXPERIMENT

A great number of the graphs from scientific and technical fields are used to depict the results of an experiment. In an experiment, the person conducting the experiment (the "experimenter") systematically changes (or manipulates) some variable, and for each change makes measurements of some other variable. The variable that is under the experimenter's control is called the **independent variable**. The variable that is measured is called the **dependent variable**.

The experimenter arranges for the independent variable to be one of a number of fixed values. These fixed values are independent of anything else and will not be altered in conducting the experiment. The measure of the dependent variable may change in some way, depending on which value of the independent variable the experimenter has set. The dependent variable depends on the prevailing value of the independent variable. The experimenter also tries to control or hold constant other variables (called **extraneous variables**) that might influence the value of the dependent variable.

Consider a hypothetical medical experiment in which patients with high blood pressure are given doses of a drug designed to treat the condition. The experimenter manipulates dose level of the drug, and dose level is the independent variable. Patients with high blood pressure are assigned at random to one of the groups in the experiment. Because of random assignment, each group starts with nearly the same average (high) blood pressure; each group will receive a different dose of the drug. Let's say that the average diastolic blood pressure for each group is about 110 millimeters of mercury. (Diastolic blood pressure is the smaller number in a blood pressure reading, representing pressure when the heart is not contracting; normal diastolic pressure is 80 millimeters of mercury or less.)

The experiment is conducted by giving each group a different dosage level of the drug, taken each day for three months. After three months, blood pressure readings for each group are again measured. We define the dependent variable to be the average diastolic blood pressure after drug therapy. If the drug is effective, blood pressure will be reduced.

Included in this experiment is a **control group** of patients who are not given the drug at all, but are given a daily dose of a **placebo**, a pill containing an inactive substance such as sugar. As you might know from studying research methods, a placebo condition controls for factors such as patients' expectations that any pill will affect them. We expect little or no reduction in blood pressure for the control group. In research design, the kind of experiment we have just described is called an **independent groups design**, because several separate groups each receive a different value (different dose) of the independent variable.

The results from this hypothetical experiment are graphed in Figure 4-2. By convention, the independent variable is always plotted against the X axis, which is labeled "drug dose" and has values of 0 (the control condition), 50, 100, 150, 200, and 250 milligrams per day. The dependent variable is always plotted against the Y axis, one measure for each value of the independent variable. The dependent variable is labeled "diastolic blood pressure after drug therapy." The graph thus shows average after-therapy blood pressure for each group of patients taking a particular dose of the drug.

We read the graph as an inverse nonlinear relationship. At 0 dose level, average blood pressure is still near 110 points (high pressure), as we expected for the control condition. The groups of patients that received 50 and 100 milligrams of drug had small reductions in blood pressure (to 102 to 105 points) and the slope of the change is shallow. At a dose of 150 milligrams, blood pressure is down to 95 points, and drops with a steep

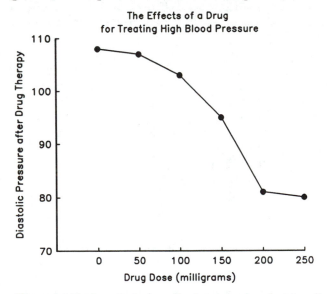

Figure 4-2 The hypothetical results of an experiment on the effects of a drug on high blood pressure.

slope to 81 points for a daily drug dose of 200 milligrams. At a dose of 250 milligrams, blood pressure is normal at 80 points.

What is the interpretation of this graph? We might like to conclude that the drug is effective in reducing blood pressure. This amounts to saying that a **cause-and-effect** relationship exists, and that increasing drug dose causes a reduction in blood pressure. If cause and effect holds, the drug may indeed be useful in clinical attempts to control high blood pressure.

Although it may seem perfectly reasonable to make a cause-and-effect interpretation of the outcome of this experiment, several criteria must be met. First, the two variables must **covary** (which they do in this case since blood pressure changes with dose level). Second, they must have a **time-order** relationship (here this is true since blood pressure changed **after** drug therapy). Finally, we must be able to rule out **alternative explanations** for the effect graphed in Figure 4-2.

This last criterion is often the hardest to show conclusively, but if an experiment is well designed it will usually hold. A major factor, for example, is whether the groups of patients were equivalent prior to the therapy. We know that each group had the same average high blood pressure, but were the groups equivalent in average age or weight, for example? The process of random assignment of patients to the various groups is intended to form equivalent groups with the same average characteristics such as weight, height, or even intelligence or anxiety level.

Further, were the groups treated equivalently? For example, were all doses of the drug administered in pills that were the same shape, size, and color? Hopefully, the experimenter controlled these other variables so that they were indeed held constant.

Questions such as these may seem overly picky, but the outcome of an experiment is not accepted as valid unless we can rule out the possibility that extraneous variables actually produced the results, rather than the independent variable that was manipulated. As we have indicated, techniques such as random assignment of subjects and holding other variables constant are ways of minimizing the influence of possible extraneous variables.

To explore the issue of causality in experiments completely would require an entire course on research design. When you interpret graphs that present the results of experiments, at the very least you should carefully read the verbal description of the experiment (including the methods or procedure used) to help decide whether a cause-and-effect interpretation is justified. If you can spot flaws in the design or logic of the experiment, this may change your interpretation of the results in the graph.

Returning to Figure 4-2, if it turns out that the experiment had no obvious flaws or alternative explanations, we could go ahead and conclude that the drug causes therapeutic changes in blood pressure. We would also conclude that a dose of at least 200 milligrams would return patients to normal, but that lower doses might have much smaller effects. Before certifying the drug for clinical use, we would want to ask other questions, especially concerning possible side effects of the drug, or adverse effects of long-term use of the drug. If you are getting the idea that the interpretation of a graph of experimental results is an ongoing and multifaceted process, you are right!

Experiments can be performed on a vast variety of variables, in the physical world as well as in the physiological, behavioral, or mental reactions of humans or animals. In the physical sciences, the independent and dependent variables are usually physical variables such as mass, amplitude, time, distance, the concentration of a chemical, the temperature of a solution, and so on.

In behavioral sciences, the variables depicted in graphs of experiments are a little more diverse. In the field of psychology, for example, some measurable aspect of behavior is usually the dependent variable in the experiment (such as the number of items answered correctly in a problem-solving task), while the independent variable might be a physical one (such as the intensity of a light in a visual perception experiment) or a procedural one (such as specific instructions given to a subject who is attempting to remember a list of words).

In high-technology fields, such as electronics, experiments may consist of evaluating the input-output relationship for machines. For example, the performance of a stereo amplifier can be graphed by plotting the strength of an output signal (the dependent variable appearing on the Y axis) versus the audio frequency (pitch) of an input signal (the independent variable on the X axis). A graph of the output of a high-quality amplifier is flat (a constant relationship) as the input frequency varies from low (bass) to high (treble). No matter what classes of variables are involved, the crux of an experiment is that an independent variable is systematically changed or manipulated while measures of a dependent variable are taken. The graph of the results invariably shows how the dependent variable changes as a function of changes in the independent variable. Figure 4-3 depicts some of the ways in which the vertical and horizontal axes of a graph can be defined and interpreted, given that the data come from an experiment.

INTERPRETING A CORRELATIONAL GRAPH

In conducting research in any field, it is always possible to make measurements of interesting variables but it may not be possible to conduct a true experiment (that is, to be able to exert control over an independent variable). For example, in research on public health, it is possible to

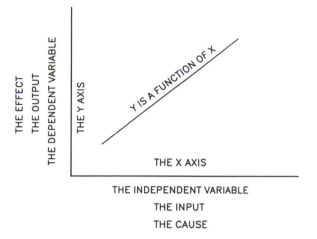

Figure 4-3 Some ways to define the axes of a graph if the graph shows the results of an experiment.

measure the amount of alcohol a woman consumes during pregnancy and to also record whether her pregnancy involved any complications or birth defects in the child she carried. But it is not possible, purely on ethical grounds, to manipulate alcohol consumption (ask a woman to drink a certain amount) to see how alcohol affects the outcome of her pregnancy. In the field of meteorology, it is possible to record average daily wind speed in a given area and to also measure average daily temperature. But it is not possible, in this case on practical grounds, to control wind speed to see its effects on temperature.

Cases such as these, where simultaneous measurements of two variables are taken, but neither variable is actually manipulated, are called **correlations**. Figure 4-4 shows a graph (a **scatterplot**) from a hypothetical correlation. Suppose that for each month of a given year in the city of Chicago, we record two measures: the number of murders committed and the number of air conditioners sold.

The scatter in the graph comes from variability in the two measurements. In a correlation, the values of both the X variable and the Y variable may vary (in a given month, both murders committed and air conditioners sold may vary). In an experiment, there is also "scatter" in the data, but only the value of the dependent variable has variability (up-down variation of data points along the Y axis), since the values of the independent variable are fixed by the experimenter. The concept of variability in data is also treated in Chapter 5 (Section D).

According to Figure 4-4, there is a direct relationship between murders and air-conditioner sales in Chicago: As sales increase, so does the number of murders. How shall we interpret this, that selling air

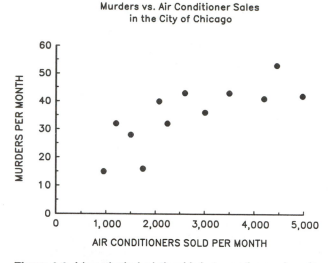

Figure 4-4 A hypothetical relationship between the number of murders committed and the number of air conditioners sold each month for 12 consecutive months in the city of Chicago.

conditioners **causes** more murders? Should we ask the mayor of the city to ban the sale of air conditioners in order to fight crime? Hopefully you agree that a cause-and-effect interpretation of this graph is not justified. All we have observed is a correlation between the two variables, that they change together in the same direction (direct relationship). Some other variable must be the cause of the changes observed. In this example, maybe the average temperature each month in Chicago governs both variables: The hotter it is, the more likely it is that murders will be committed and the greater the sales of air conditioners. For many observed correlations, some unmeasured (often unknown) "third variable" is the cause of the relationship depicted in a graph. In general, with data points derived from a correlation, we must look to factors outside of those depicted on the graph to completely interpret the graph.

INTERPRETING A TIME SERIES GRAPH

One of the most frequently encountered types of graphs is a time series graph, in which the variable on the X axis is the passage of time. In such a graph, measurements of some kind are made every minute, hour, day or year, and the graph shows how the measurement (plotted on the Y axis) changes as time (plotted on the X axis) goes by. A huge number of social and economic measures, such as unemployment rate, ownership of homes, gross national product, and the incidence of diseases, are routinely

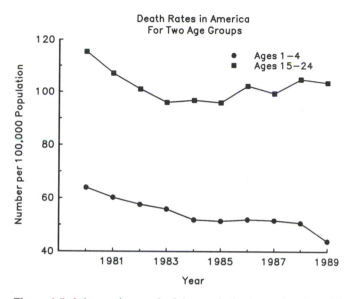

Figure 4-5 A time series graph of the yearly death rate (number of deaths per 100,000 population) for Americans aged 1 to 4 years (closed circles) and 15 to 24 years (closed squares) for the period 1980 to 1989. *Source of data*: National Center for Health Statistics (1990).

graphed as a function of time. How do we approach the interpretation of such graphs?

First we note that the passage of time is not at all like the independent variable in an experiment. All things change over time, but we do not think of time as causing the changes. Instead, other variables that are modulated with time are responsible for changes in the measured variable.

Second, many variables change in cycles with the passage of time. As a trivial example, consider that temperature and precipitation show cyclical changes over the year with the passing seasons. In economics, it is widely thought that variables such as interest rates and business volume follow cyclical patterns. It follows that any time variation we observe in any variable could potentially have a cyclical component.

Third, many variables show natural, random fluctuation over time. The value of a variable may be especially high or low because of random, unpredictable factors. The temperature may follow a cyclical pattern over the year, but from day to day, it can vary considerably (as a Chicago resident, I will testify to that fact).

The most basic time series graph simply plots measures of a selected variable over time to display potential positive or negative trends in the value of the variable. As an example, in Figure 4-5 we plot a (somewhat grim) statistic, the yearly death rate (number of deaths per 100,000 population) among Americans aged 1 to 4 years and 15 to 24 years during the 1980s (National Center for Health Statistics, 1990). For young children, the time series relationship is inverse, with the death rate declining over the decade. For the older group, the relationship is U-shaped, with high death rates at the beginning and end of the decade.

How do we interpret this graph? For young children, it seems plausible that improvements in child health care and nutrition may have contributed to the gradual decrease in the death rate. But what about the data for people from 15 to 24 years old? As a person in this age group, should you be alarmed by the clear increase in the death rate from 1985 to 1989? And what was responsible for the sharp decline in the death rate from 1980 to 1983? Does the graph depict genuine trends, a random down-up fluctuation or is it part of a long-term cycle? It is difficult to be conclusive about directional trends in any variable, such as death rate, that is subject to a good deal of fluctuation over the long term. Further, we can imagine many factors (incidence of serious accidents, increases or decreases in the occurrence of mortal illness such as leukemia or cancer, interpersonal violence and crime) that might affect a variable such as death rate among young Americans.

Not all simple time series graphs are so hard to interpret. Figure 4-1 shows us a distance versus time relationship between earth and a comet that had a very unambiguous interpretation. Time trends in variables such as population growth or unemployment rate are often similarly clear cut.

There is a second type of time series graph that involves plots of measurements taken before and after a specific event or manipulation. This is called an **interrupted time series graph.** As a hypothetical example, suppose that the mayor of Chicago is concerned about the large number of pedestrians who jaywalk rather than using crosswalks. The fine for jaywalking has been $5, but the mayor has the fine raised to $50 on January 1. The issue is, if the increase in the fine has lead to fewer arrests for jaywalking.

Figure 4-6 gives the evidence on this issue, plotting weekly total arrests for jaywalking in Chicago for six weeks prior to and six weeks after the fine became $50. After the new fine, arrests for jaywalking have declined gradually but steadily. Before interpreting this to mean that the $50 fine has caused less jaywalking, we need to explore alternative explanations. Suppose the hoards of Christmas shoppers artificially inflated the pre-$50 fine data, while severe January weather kept people off the streets and reduced the post-$50 fine data. Perhaps New Year's resolutions to not jaywalk reduced arrests independent of the fine increase.

The point here is that in an interrupted time series, many variables may effect a change observed on a graph. As with a correlation, the change

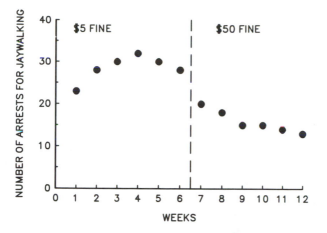

Figure 4-6 A hypothetical interrupted time series graph of the number of arrests for jaywalking during the four weeks prior to and after an increase in the fine for jaywalking from $5 to $50 in the city of Chicago.

may be due to a variable other than the event or manipulation that preceded the effect. Both correlations and time series lack control conditions in which the independent variable has a zero value. Without a control condition, there is no baseline for comparison with effects due to the independent variable. In our example, we might want to compare Chicago jaywalking arrests with those in Milwaukee during the same period (assuming Milwaukee didn't increase the fine).

PRACTICE IN GRAPHIC INTERPRETATION

This completes our discussion of data sources and graph interpretation. Do you feel that your mental knives are sharp enough to tackle some interpretive problems? Are you sharp, sharp, sharp when it comes to cause and effect, extraneous variables and alternative explanations for the relationships shown in graphs? Let's sharpen our knives on the following cases.

Case 1 (Psychology): Practice, Practice, Practice

Herman Ebbinghaus (1885; 1964) was one of the first psychologists to study human memory in a systematic way. He concentrated on the memorization of lists of nonsense syllables which are three-letter consonant-vowel-consonant sequences (examples, ZAL, SED, or LIR). Ebbinghaus invented the nonsense syllable to serve as "pure" verbal material without any meaning or other association. Ebbinghaus, who served as the sole subject for his experiments, would practice a list of nonsense syllables by repeating it for a fixed number of times. He was extremely disciplined in his practice—he always recited a list of syllables at 150 items per minute (regulated by the ticking of a metronome), he tried to avoid accents or stresses in pronouncing the nonsense syllables, he practiced a given list at the same time each day, and he exercised extreme concentration as he repeated the essentially meaningless syllable sequences over and over.

Figure 4-7 depicts some of his most notable data. On one day, various series were repeated (i.e., practiced) a certain number of times, ranging from 8 to 64 times. (For any list, even 64 repetitions was not enough to learn the list "by heart," without errors.) On the next day, Ebbinghaus measured the number of additional repetitions it took to finally learn each list perfectly without any mistakes. The graph plots a nearly linear decrease in time to learn a list of syllables as a function of increasing number of initial repetitions of the list. The greater the number of first-day repetitions (amount of practice), the less time required on the second day to master a list of syllables. How do we interpret this graph? Write down your thoughts about Figure 4-7, then read the following.

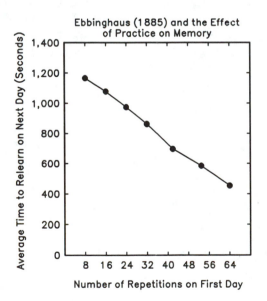

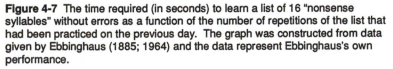

Figure 4-7 The time required (in seconds) to learn a list of 16 "nonsense syllables" without errors as a function of the number of repetitions of the list that had been practiced on the previous day. The graph was constructed from data given by Ebbinghaus (1885; 1964) and the data represent Ebbinghaus's own performance.

Interpretation

You should have noted, first of all, that the graph depicts the results of an experiment. The independent variable is the number of repetitions of each list on the first day of memorization. The dependent variable is the time required to learn a list perfectly on the second day (since Ebbinghaus repeated his lists at a constant number of syllables per minute, time translates directly into additional repetitions on the second day). In research methods, this type of experiment is called a **within-subjects design**. This means that each subject (here, just Ebbinghaus) is exposed to all values of the independent variable (the number of repetitions of a list). Usually, the values of the independent variable are experienced by a subject in random order so that extraneous effects of the sequencing of values of the variable are balanced out.

Can we interpret a cause-and-effect relationship between the variables? The short answer is yes, given that Ebbinghaus was a careful experimenter who tried hard to control extraneous variables. We can say therefore that increasing the amount of practice in a memory task causes a progressive improvement in learning, and less and less time is required to master a list of items during a second learning session.

The most remarkable thing about this effect of practice is its **linear** nature (Baddeley, 1990) as is evident in Figure 4-7. Any given increase in practice (from 8 to 16 repetitions or from 16 to 24 repetitions), produces a fixed decrease in the time required to master a list of syllables (of about 100 seconds). A linear relation implies that some relatively simple process or mechanism is responsible for the experimental effect. The linear effect of practice in turn is probably related to the "pure" nature of nonsense syllables. Since nonsense syllables have no verbal associations or meanings, we get a pure measure of the effect of practice.

In sum, you should have interpreted the graph as representing the results of an experiment, as probably being a cause-and-effect relationship, and you should have taken note of the importance of the linear relationship. Any further interpretation of this graph would take us into an open-ended discussion of the nature of human memory systems, such as can be found in an undergraduate course on the experimental psychology of learning and memory.

Case 2 (Sociology): Knowledge and Nuclear Power

Figure 4-8 is a graph of data collected in a survey of knowledge about and attitudes toward nuclear power. The survey was conducted for the Columbia University School of Engineering; the data were analyzed in an article by Cole and Fiorentine (1988). The graph shows the relationship between level of knowledge about nuclear energy and level of favorable

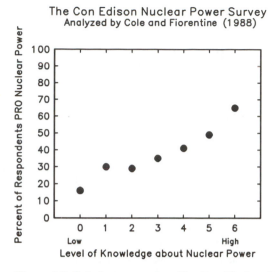

Figure 4-8 Data from a survey polling the attitudes of New York City area residents about the use of nuclear power by Consolidated Edison.   The graph plots the proportion of respondents favoring nuclear power as a function of their level of knowledge about nuclear power.

attitude toward nuclear energy.   Knowledge level was assessed with questions like: "Can a nuclear plant blow up like an atomic bomb?" Attitude was assessed with questions like: "In general, do you favor or oppose the building of more nuclear power plants in the United States?" All respondents to the survey were categorized according to how many knowledge questions they got right (Knowledge Index), and Cole and Fiorentine tabulated the percent of respondents in a given knowledge category who favored the use of nuclear power. These are the variables that are plotted in the graph.

We clearly read a direct relationship between the variables: The more knowledge one has about nuclear power, the more likely one is to favor nuclear power. Now go ahead and note your interpretation of this graph.

Interpretation

You should have said that the graph plots the results of a correlation, since both knowledge and attitude are being measured at the same time. Does increased knowledge about nuclear power cause more favorable attitudes toward nuclear power? The owners of a power company would like it to be true that knowledge causes attitude changes, since "that sees opposition to nuclear power as a result of ignorance of the 'real' risks." (Cole and Fiorentine, 1988, p. 308) Opposition to nuclear power could therefore be reduced simply by increasing the amount of education about nuclear power that the public is exposed to.

But since the graph presents data from a correlation, there is no sure way of knowing that cause and effect run in a specific direction. As Cole and Fiorentine point out, perhaps knowledge doesn't influence attitude. Instead the reverse may be true: Perhaps attitudes influence knowledge. If you start out with a negative attitude toward nuclear power, you might be inclined to state that nuclear plants can blow up like atomic bombs because this incorrect answer is consistent with your attitude. Cole and Fiorentine go on to cite a third relevant variable—fear of nuclear power. Perhaps people are opposed to nuclear energy because they are afraid of it, and the less they know, the more afraid they are likely to be.

In summary, this is a classic case of the difficulty of conclusively interpreting a graph showing a correlation between two variables. What might seem to be cause and effect is not necessarily so.

Case 3 (Education): The Scholastic Aptitude Test—Males, Females, and Math

Figure 4-9 was constructed from data in *The World Almanac and Book of Facts (1991)*. (The original source is the College Entrance Examination Board.) The graph shows the mean score on the mathematics section of the Scholastic Aptitude Test obtained by college-bound high

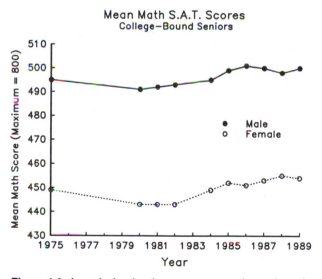

Figure 4-9 A graph showing the mean score on the math section of the S.A.T. for college-bound males (filled circles) and females (open circles) from 1975 to 1989.

school seniors, male and female, from 1975 to 1989. The graph depicts a decline in math scores from 1975 to 1980, but a general increase in scores since 1980. For every year shown, the average score for females has been lower than that for males by 40 to 50 points.

How do you interpret the data?

Interpretation

You probably see that this is a classic time series graph that simply plots measures over time. The graph clearly documents an overall improvement in math scores since 1980 and a clear male-female difference. What do these trends in the data mean?

Without considerable additional information, there is not much we can say definitely about the meaning of this graph, and in fact it raises many more questions than it answers. Are students getting better at math since 1980, perhaps due to improvements in math education, or is the test simply getting easier? Is the pool of test takers getting smaller or larger, and how would that influence the average math score? Is there a "sex difference" in mathematical ability, or does the difference between males and females reflect biases built into the system of math education or biases that appear in the actual S.A.T. exam? Because the data on the graph are simply "discovered" out in the real world, many possible factors may have caused the measurements to be what they are.

Data such as those in Figure 4-9 are provocative and may lead us to pose research problems and examine practical issues. On the basis of these data, for example, one might want to conduct experiments on the effects of different techniques of math instruction on measured math ability. Or one might be led to conduct experiments to test the validity of male-female differences in math scores. While we cannot give a firm interpretation of data such as those in Figure 4-9, the graph functions to stimulate thinking on the measurements that are plotted.

CHAPTER SUMMARY

The aim of graph interpretation is to understand the significance, the cause, and the implications of the data plotted on the graph. Reading a graph involves identifying a data relationship; interpreting a graph is an effort to comprehend the meaning of the relationship. Reading a graph is a closed-end process that can be completed, while graph interpretation is an open-ended process subject to continuing revision. The key to successful graph interpretation is to understand the source of the data plotted in the graph. Many graphs are based on data from experiments, in which an experimenter manipulates one variable (the independent variable) and measures another variable (the dependent variable). In a well-controlled experiment, we can interpret a graphic relationship as indicating that the independent variable causes changes in the dependent variable. Graphs can also depict data from correlations, in which measures of values of two variables are taken at the same time. The graph of a correlation may show a clear relationship between two variables, but since no experimental manipulation is made, it is not usually possible to infer cause and effect. The same problem occurs in a time series graph, where a sequence of measures of a variable is made. In correlations and time series graphs, data relationships may be governed by alternative factors (extraneous variables) that were not measured. In all three cases—experiments, correlations, and time series—graphic interpretation should proceed with intellectual skepticism that is continually seeking alternative explanations for the data relationship depicted in a graph.

| 5 | Special Topics |

Plan of the Chapter

PLAN OF THE CHAPTER

The final chapter of this book is optional. At this point, if you need to plunge into a textbook or report full of graphs, have no fear, you will understand what you see. But you might encounter something on a graph that you haven't seen before, such as a log scale, on which numbers are spaced by the value of their logarithm, or error bars sticking out of the top and bottom of a data point (like little antennas). You might wonder what these things mean. Chapter 5 covers such details.

This chapter is made up of a series of brief discussions and explanations of special topics that help round out a person's basic graph literacy. Chapter 5 can help with a course on research methods, in which you find graphs that represent more than two variables and have multiple data curves. In a course on statistics, the chapter can shed some light on the concept of the frequency distribution or the best-fitting regression line. If you are interested in high-tech aspects of graphs, the chapter concludes with a short description of what computer graphics programs can do.

Even if you choose to skip these special sections, please read the concluding comments at the end of the chapter.

A. THE FREQUENCY DISTRIBUTION

A **frequency distribution** is a special type of X-Y data set that is useful in statistics and research design. To make a frequency distribution, we start with a set of measurements of a particular variable (X). The measurements will typically not all have the exact same numerical value, but will vary over a range of values. For each measured value of X, we tally up how many times (how frequently) that value has appeared in the set. The Y variable is defined as the frequency of a given X value.

As an example, suppose that in a class of 100 students, we determine how many students have various numbers of siblings. The number of siblings (say 0 to 6) is the X variable (the value of X can be 0 siblings, 1 sibling, 2 siblings, and so on). The number of students having a certain number of siblings is the Y variable. A data set from the group of students might look like that in Table 5-1. Table 5-1 is a **discrete frequency distribution**, meaning that the X values are whole numbers (integers). The X variable is not continuous; it is not possible to have 2.32 siblings, for example.

A graph of this frequency distribution is shown in Figure 5-1. From Figure 5-1, we can tell that most students have 0, 1, or 2 siblings, and that 2 siblings is the most frequent value of the X variable.

Table 5-1

A frequency distribution of number of siblings among a group of 50 students.

Number of Siblings (X)	Frequency or Number of Cases (Y)
0	16
1	22
2	30
3	18
4	4
5	6
6	4
	100 cases **Total**

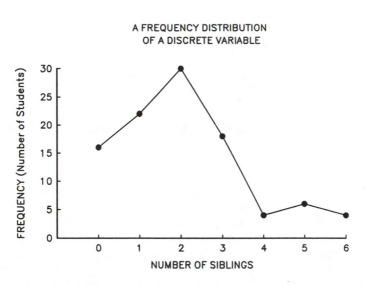

A FREQUENCY DISTRIBUTION
OF A DISCRETE VARIABLE

Figure 5-1 A frequency distribution of a discrete variable, the number of siblings had by each of a class of 100 students.

Now suppose that we measure the height (in feet) of each of the 100 students in the group. Height is a variable that has very fine gradations in level (called a **continuous variable**). We therefore measure the heights of all students, but then group the measured values into class intervals. For example, if a student is 4 feet, 8 inches (4.66 feet), we tally a case in the interval 4.5 to 4.75 feet. Lumping values that are actually different into the same class intervals causes some loss of precision, but the use of class intervals is a convenient way to deal with variables, like height, weight, or income, that are continuous.

Now consider the purpose for making a graph of a frequency distribution of data. One function of a frequency distribution is to illustrate where along the range of the X variable cases are concentrated. For example, consider the height data from Table 5-2 that are graphed in Figure 5-2. We see at a glance that the peak of the distribution (also

Table 5-2

A frequency distribution of height among a group of 50 students.

Height in feet (X)	Frequency or number of cases (Y)
4.0-4.49	3
4.5-4.99	11
5.0-5.49	25
5.5-5.99	33
6.0-6.49	23
6.5-7.0	5
	100 cases **Total**

A FREQUENCY DISTRIBUTION
OF A CONTINUOUS VARIABLE

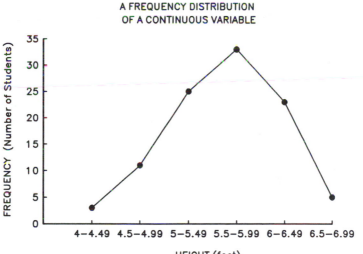

HEIGHT (feet)

Figure 5-2 A frequency distribution of a continuous variable, the height of each of a class of 100 students. Height measures are put into class intervals of 0.5 feet in size.

known as the **mode** or 'most frequent case') lies in the interval 5.5 to 5.99 feet. In this distribution, which is approximately **normal** or **bell-shaped**, the peak of the distribution will be very close to the arithmetic average (**mean**) of all values of X. The first function of a frequency distribution is to illustrate the central tendency of the data set.

The second function of a frequency distribution is to show how much variability there is in a data set. From Figure 5-2, for example, we see that the data may range from 4.5 feet to 6.75 feet even though most of the cases lie between 5.25 and 6 feet. In general, all measurements exhibit variability, which can also be thought of as error. All real-world measurements of anything whatsoever are subject to **error**. The graph of a frequency distribution gives us an important visual representation of the amount of error in a measurement. Furthermore, the frequency distribution helps us visualize a famous statistical measure of variability in data, the **standard deviation**.

If values in a distribution are concentrated around the mean value, the standard deviation of the distribution will have a small numerical value. If the values are spread out, the standard deviation of the distribution will be larger. But so long as a frequency distribution is symmetrical (the same shape above and below the mean value) and also bell-shaped (a **normal distribution**), then a **constant 67 percent of all cases** of the X variable will lie in the window defined by the **mean plus or minus one standard deviation**. This concept is illustrated in Figure 5-3.

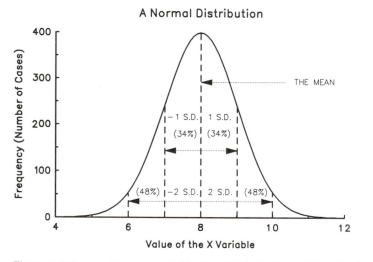

Figure 5-3 A normal frequency distribution. This distribution is idealized—the X variable is continuous, there are no class intervals, and a very large number of scores are represented. Marked on the X axis are the mean of the distribution, and plus and minus one and two standard deviations from the mean. The standard deviations encompass fixed percentages of all scores, as shown on the figure.

Now you know why they call it the **standard** deviation. Step away from the mean one standard deviation, and you enclose a standard 67 percent of the distribution. Two standard deviations above and below the mean value will take in slightly more than 95 percent of all cases, and three standard deviations will encompass more than 99 percent of all values of X. For any given number of standard deviations away from the mean, a constant percent of cases in the distribution will be included. These properties of the standard deviation are the foundation of inferential statistics and statistical decision making (a story that is beyond the scope and space of this book).

It is also the case that the normal distribution is important because it describes the distribution of a large number of behavioral, biological, or physical variables. In human beings, for example, measures such as height, weight, and the intelligence quotient (I.Q.) tend to have normal distributions.

Not all frequency distributions are as nice and symmetrical as the one in Figure 5-3. A distribution may turn out to be lopsided, with more of the cases on one side of the peak frequency than on the other. We refer to such unbalanced distributions as **skewed**. Figure 5-4 shows a graph of a skewed distribution, the frequency distribution of income among a number of families. Real distributions of income are skewed toward higher levels. A relatively few families are well-off (income levels above the peak frequency) and lots of families are of modest means (income levels below the peak frequency).

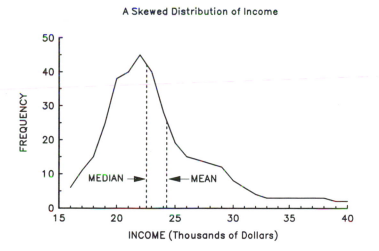

Figure 5-4 An idealized frequency distribution of family income, which is a skewed distribution. Marked on the X axis are the mean, which divides the total dollars of income in the distribution into two equal halves, and the median, which divides all income levels in the distribution into two equal halves.

There are two arrows along the horizontal axis in Figure 5-4. The one on the right marks the **mean** value or simple arithmetic average of family incomes. The mean specifies how many dollars of income there are per family.

The arrow on the left points to the **median** value of family income. The median is defined as the income level which 50 percent of the total cases lie above, and 50 percent lie below. In Figure 5-4 the median divides up the distribution so that one half of the families lie above the median income level, and one half lie below it. When a frequency distribution is skewed, as income level is here, it is always better to figure the central tendency or "average" based on the median rather than the mean. The presence of extreme values above the mean tends to inflate the mean or arithmetic average. It is more useful to know the income that divides all households, so the federal government routinely reports "average" family income as the median income.

B. THE BAR GRAPH

The bar graph is an alternative form for a two-dimensional graph that is used a lot in business publications and newspapers. Recall that in Chapter 1, we state that if a data set is qualitative (the X variable is a label, the Y variable is a number), we should not construct a line graph; instead, a bar graph should be used to the plot data. A bar graph is shown in Figure 5-5.

MOTOR VEHICLE PRODUCTION, 1989

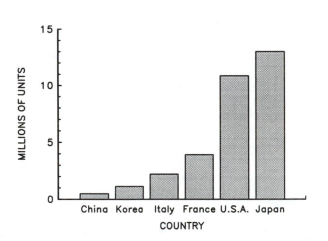

Figure 5-5 A bar graph of qualitative data set. Source of data: *The World Almanac and Book of Facts, 1991.*

Values of a qualitative variable are placed on the horizontal axis, so that the value of the quantitative variable is represented by the height of the bars.

Bar graphs can be used to graph frequency distributions (in that case, the graph is called a **histogram**). Figure 5-6 shows the frequency distribution of Figure 5-1 as a histogram. This is an application of the bar graph in which both variables are quantitative.

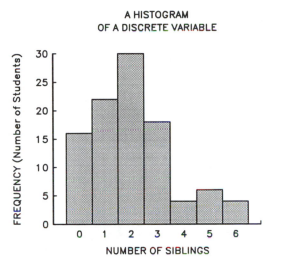

A HISTOGRAM
OF A DISCRETE VARIABLE

Figure 5-6 A histogram of the frequency distribution from Figure 5-1 (Table 5-1).

C. LOGARITHMIC GRAPHS

Some line graphs that you may encounter will have a **logarithmic scale** on one or both axes. The logarithm of a number is simply a power or exponent to which another number is raised. Consider this: You already know that 10 raised to the second power equals 100, and that 10 raised to the third power equals 1,000. We define the log of 100 to be equal to 2. That is, 2 is the power (or logarithm) to which 10 (the base) must be raised to equal 100. Since 10 raised to the third power (10 x 10 x 10) equals 1,000, it follows that the log of 1,000 should equal 3. What is the log of 10,000? That's right, 4 is the log of 10,000. That's because 10 raised to the fourth power equals 10,000.

Table 5-3 lists the logarithms of numbers from 10 to 100. The log of a 20, for example, is 1.3 since 10 raised to that power equals 20. Now let's make a graph of the table to find out what form of relationship exists between numbers and their logarithms. In Figure 5-7, we plot the numbers 10 through 100 on the horizontal axis, and the log of the numbers on the vertical axis.

Table 5-3

Number	Log (Base of 10)
10	1.0
20	1.3
30	1.48
40	1.6
50	1.7
60	1.78
70	1.84
80	1.9
90	1.95
100	2.0

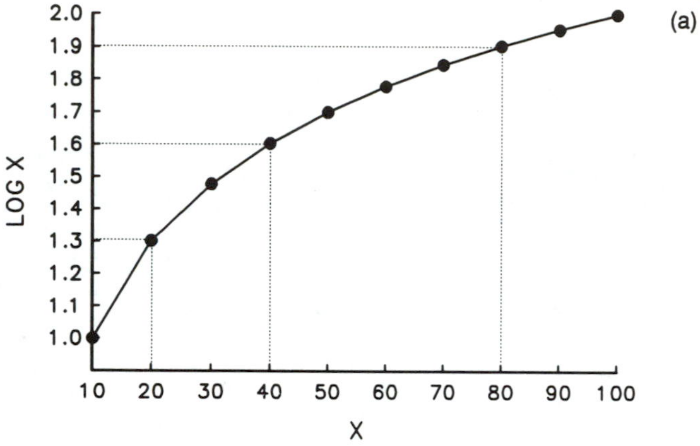

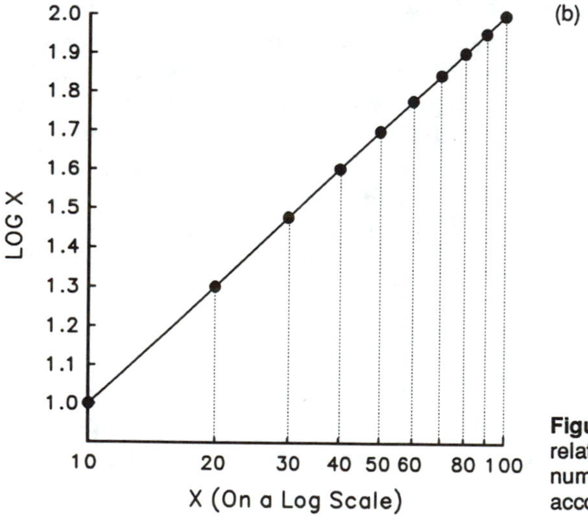

Figure 5-7 (a): A graph of the relationship $Y = LOG\ X$. (b): The numbers on the x axis are spaced according to their logarithms.

As you can see, the log relationship is a direct decelerated one. The figure also illustrates a critical property of logarithms: When numbers change by a constant multiplier or factor, then the corresponding logarithms change by a constant amount. For instance, when we go from 20 to 40 to 80 (a progressive doubling of numbers), the logs go from 1.3 to 1.6 to 1.9 (a progressive addition of the constant amount 0.3). To put this another way, if a variable is changing by a constant factor or percent, its logarithm is changing by a constant amount.

To create a log scale of the numbers from 10 to 100, picture "straightening out" the curve in Figure 5-7 by pulling the end of the line up and allowing the spacing between numbers to change so that the relationship between the Y value (the log) and the X value (the number) is linear. This is shown in the other panel of Figure 5-7.

Now let's picture a graph with a **log scale** on the Y axis and the passage of time on the X scale, a **semilog time series graph**. This semilog graph is ideal for plotting percentage or multiplicative changes that occur over time, such as the effect of compound interest. Compound interest works like this: A sum of money is invested at a certain interest rate, and total proceeds are allowed to accumulate from year to year. The growth of the total dollars accumulated over years is shown in Figure 5-8. As the figure shows, if we plot dollars versus time, the relationship is

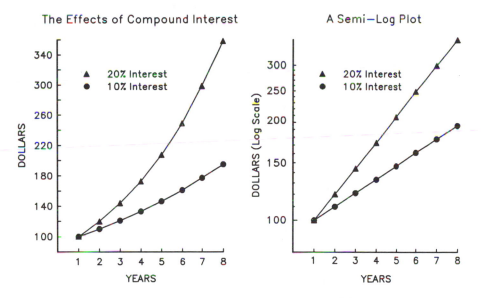

Figure 5-8 The use of the semilogarithmic graph to show rate of change. A constant rate of increase of a variable (such as 10 percent yearly interest rate on an amount of money) produces a constant change in the log value of the variable (log money increases by a constant amount). On a plot of log dollar versus time, a constant rate of increase produces a straight line (a constant slope).

accelerated. But if we plot log dollars versus time, we get a straight line. The slope of the semilog plot is an indicator of the rate at which the money grows. If an amount of money doubles every five years, for example, this means that log money increases by a constant amount every five years (on a plot of log dollars versus years, the growth of money is a straight line). The higher the slope, the higher the rate of increase (the percent interest on the money, for example). In general, a semilog plot is an excellent way of visualizing rate of change, since a constant rate produces a straight line on a semilog graph.

As one additional example of the use of a semilog time series graph, consider plotting the way a person's salary changes over the years. If you make a practice of plotting the logarithm of your salary over years, you will have a very clear idea of how you are doing, financially speaking, in your job. Figure 5-9 shows the salary history of three workers as plots of log dollars versus years of employment. In Figure 5-9, the slope of the top salary curve is increasing over time. Since we are plotting log dollars of salary, this indicates that the percentage yearly raise is increasing. Not only is the worker getting paid more every year, but his rate of increase is itself increasing. Conversely, if the log salary slope gets more shallow, then you are losing ground—your rate of increase (percentage raise) is getting smaller. A log salary curve with a constant slope indicates a steady rate of change, yearly raises of the same percentage. Making a graph of

Figure 5-9 Three semilog salary histories: percentage raise increasing (up and coming), percentage raise constant (holding steady), and percentage raise decreasing (losing ground).

your log salary over the years is a clear visual indicator of whether you are "up and coming," "holding steady," or "losing ground."

One other useful property of log scales is their ability to represent a very large range of numbers. With a log base 10 scale, every unit change in the log of a number is equivalent to multiplying it by 10. A numerical range of, for example, 1 to 1 million can be represented with a log scale of just 6 units, since 10 raised to the sixth power equals a million.

In my own field of experimental psychology, measures of visual sensitivity are often expressed with log scales because the range of sensitivity is so large. For example, going from a darkened room to a fully-lighted one reduces light sensitivity by a factor of more than 10,000. Rather than trying to represent the range of 10,000 on a vertical axis (where the dependent variable is light sensitivity), we use a log light sensitivity scale that runs from 0 to 4 (log of 1 is 0, log 10,000 is 4). In the physical sciences, too, many variables in the real world can change over enormous ranges which necessitates the use of logarithmic axis scales.

D. STATISTICAL ASPECTS OF GRAPHS

In this section, we briefly describe two statistical features which may appear in a line graph: **error bars** attached to data points on a graph and a **best-fitting line** that is drawn through a set of data points.

Error bars are lines that are drawn vertically above and below a data point in a graph, as depicted in Figure 5-10. Error bars are used to give

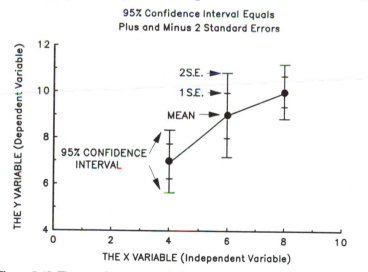

Figure 5-10 The use of error bars to indicate the 95 percent confidence interval for a data point. The value of the X variable is fixed (it is the independent variable in a functional experiment) and the value of the Y variable has varied over a set of measurements.

a visual indication of the variability associated with the measurement of the Y variable that the data point represents.

Error bars are included with data points because the vast majority of graphs in business, scientific, and technical fields are in fact used to display average values of a variable. In an experiment, for example, we make many measurements of the dependent variable and then graph the **mean value** (a plotted point is an average measure from a group of subjects, for example, or from across some number of experimental trials) for various values of the independent variable. Any mean value of the dependent variable based on a sample of data is only an estimate of the true mean value of the variable in question. (If I measure the mean height of a class of students, I get an estimate of the average height of all human beings.) Further, different measured means will also vary. (If I measure the mean height in ten different classes of students, I will get ten different estimates.) When we plot the mean of a Y variable on a graph, error bars are used to indicate a range of values wherein the true mean of the Y variable is likely to be.

The size of the error bars on a graph is usually calculated with a statistic known as the **standard error of the mean**. The standard error of a mean is an estimate of the standard deviation of the underlying distribution of all possible measured means. The standard error of a mean is computed by taking the standard deviation of the sample and dividing by the square root of the sample size.

The **95 percent confidence interval** for a sample mean is a range of values of above and below that mean within which one can be 95 percent sure that the true mean lies. This interval is equal to the sample mean plus or minus 2 standard errors, as indicated in Figure 5-10. Marking off 2 standard errors above and below a sample mean with error bars visually defines a range of Y values that includes 95 percent of the possible values of the true mean. (This is so because the distribution of sample means is normal, as in Figure 5-3.) We are 95 percent sure that the true mean— which the data point is an estimate of—lies within the range between the short horizontal ends of the error bars. (There is still a 5 percent chance that the true mean of the variable lies outside of the range since this is the nature of the normal distribution and statistical variability.)

Occasionally you may find a graph with error bars based simply on the standard deviation of a mean value. This is a mistake since the standard deviation only indicates variability among the measurements from which the mean was computed. Error bars based on the standard deviation are not useful, since they indicate nothing about the 95 percent confidence interval. Only error bars based on plus or minus 2 standard errors of the mean define a range of values where the true mean is likely to be found.

We now turn to the concept of a **best-fitting line**. We have seen several examples of a type of graph called a scatterplot that is designed to display a relationship between two measured (but not manipulated) variables (e.g., the murder rate versus air-conditioner sales example, Figure 4-4). Figure 5-11 shows a hypothetical scatterplot of data, displaying the relationship between the peak adult weight of a person and age at which they died. Plotted on the graph are life span versus weight data for a sample of ten individuals.

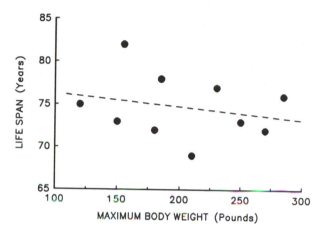

Figure 5-11 A scatter diagram of data from a correlation. Each of the data points represents the maximum body weight and the life span of a person.

The data in Figure 5-11 do not, at first glance, show any strong relationship between the variables; the data points are quite scattered, and life span seems to be uncorrelated with how much a person weighed in his or her life. With such an ambiguous case, we need to resort to statistical analysis to tell us if a trend is actually present or not.

One statistic that is used to indicate the strength of a relationship is called the **product moment** (or **Pearson**) **correlation coefficient**, which is symbolized by the letter r. The r statistic varies in value from +1.0 to -1.0. It is based on a **linear model**, meaning that the statistic has a maximum value if there is a perfect straight-line relationship between the X variable and the Y variable in a graph. If two variables have a true linear relationship with a positive slope (direct relationship), then $r = +1.0$. If the true linear relationship has a negative slope (inverse relationship), then $r = -1.0$. Any relationship that deviates at all from a straight line (that is, has some scatter in the data points) will have a value of r somewhere between +1 and -1. If $r = 0$, then the relationship between X and Y is all scatter, with no tendency for the points to form a straight line.

It is also possible to calculate a straight line to best represent any linear trend in the data. This line is called a **regression line**. A statistical regression analysis calculates the slope of the trend line and the **intercept** (where the line crosses the vertical Y axis). We do not describe how to calculate an actual regression line; for this you should consult a statistics text. We note that a regression line is a line from which the data points deviate the least; a regression line makes the best possible prediction of the values of the Y variables, given the values of the X variable. (In particular, the line minimizes the sum of squared deviations of the Y variable above and below the line, as you can learn from your statistics or research design text.) Sometimes the regression line is called a **best-fitting line** in light of the concept of minimizing deviations from the line.

In Figure 5-11 we have drawn in a best-fitting regression line for the life span and weight data. The slope is negative, meaning that increasing body weight reduces expected life span. In real research reports, it is common to find regression lines plotted that represent linear trends in data from experiments, correlations, or time series.

E. GRAPHS WITH TWO INDEPENDENT VARIABLES:
MAIN EFFECTS AND INTERACTIONS

An experiment can be used to investigate effects of more than just one independent variable at a time. A **complex experiment** can be designed in which two independent variables are combined in any experimental condition. For example, suppose we experiment to determine the effectiveness of a drug. We measure percent of patients cured by the drug (the dependent variable) for the independent variables of drug dose (low versus high) and duration of therapy (short versus long). Dose size and length of therapy can be put into **factorial combinations**, where each level of one variable is paired with each level of the second variable. This is shown schematically in Figure 5-12. This particular case is a so-called 2-by-2 design. (With two independent variables, there can be any number of levels of each variable, not just two levels. The 2 x 2 experiment is simply the smallest possible factorial experiment.)

When we graph the results of a complex (factorial) experiment, one of the two independent variables is designated as the **parameter** for the data lines that are plotted. This is illustrated in Figure 5-13. We plot percent of patients cured (the dependent variable) versus dose level (which we label the A variable). The B variable, duration of therapy, is said to be the parameter for the two data lines.

A 2−BY−2 FACTORIAL EXPERIMENT

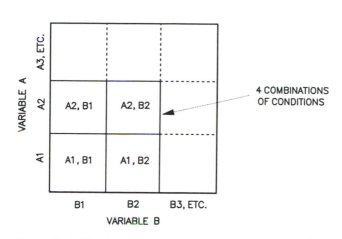

Figure 5-12 The design of a 2-by-2 factorial experiment involving two levels of some variable A and two levels of some variable B.

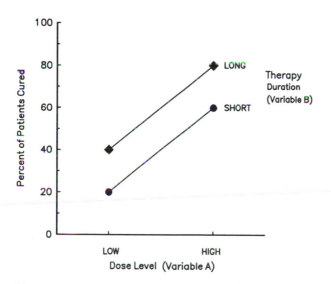

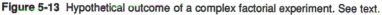

Figure 5-13 Hypothetical outcome of a complex factorial experiment. See text.

From this graph, we can read two separate **main effects** (effect of an independent variable). First, since both data lines slope up, we know that the drug is more effective at the higher dose level. Second, since the data lines are separated vertically, we know that increasing the duration of therapy also improves the drug's effectiveness (more patients are cured with the longer therapy).

Let's look at a slightly different experimental outcome, as shown in Figure 5-14. Here the data lines slope up, but they aren't parallel. We have what is called an **interaction** between the variables: The effect of drug dose (A) is influenced by the duration of therapy (B). In particular, increasing therapy duration is more effective at the low-dose level than at the high one (the two lines are separated by a greater amount at the low-dose level), although the combination of high dose and long therapy still produces the greatest percent of patients cured.

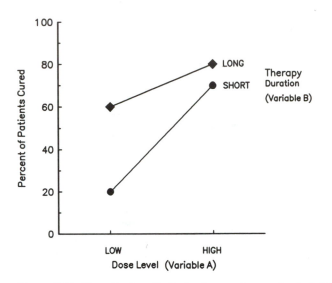

Figure 5-14 Alternative hypothetical outcome of a complex factorial experiment.

Whenever we conduct a complex experiment with two independent variables, we recognize that a total of three experimental effects can be obtained. There are two separate main effects (A, B) and a possible interaction (symbolized as A x B). Sometimes an experiment produces no main effects and all interaction. Consider giving high or low drug doses to patients who are either men or women. The hypothetical results are graphed in Figure 5-15. Here the effect of the drug is seen to be completely different, depending on whether you are a man or a woman. For men, increasing dose level increases the percent of patients cured. For women, the effect is exactly opposite. If we neglect the sex of patients, and average the results for men and women, we conclude there is no overall drug effect, as shown by the faint dotted line in the graph. If we average across drug dose, we also conclude that there is no overall effect of the sex of the patient (the average percent cured for men is the same as for women). The only factor that is significant here is the interaction. This tells us

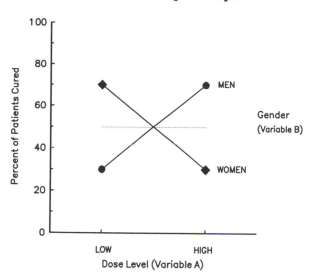

Figure 5-15 Hypothetical outcome of a complex factorial experiment—no main effects, all interaction.

that the effect of the drug is completely different, depending upon whether the patient is a man or a woman.

There is some challenging fun to be had from reading graphs of complex experiments and confirming conclusions by **replotting** the data, switching the roles of the A and B variables on the graph. From a visual point of view, an interaction can be spotted by the fact that data lines for the parameter variable are not parallel (same slope). But to determine whether separate main effects are present for both A and B variables, it is most informative to replot the data. For example, in Figure 5-16, we replot two separate sets of results from complex experiments. The trick

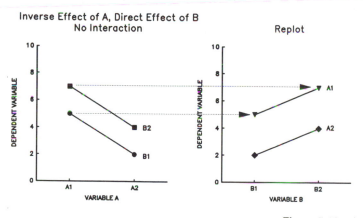

Figure 5-16 *(continues)*

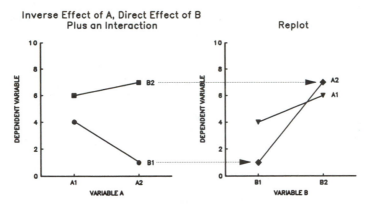

Figure 5-16 Examples of replotting outcomes of complex factorial experiments. The roles of the A and B variables in the graph are reversed.

to replotting the data is to move the Y axis values for the B variable (the parameter) in the original graph over to the X axis (main independent variable) on the new graph. For example, to get a replotted data line with A1 as a parameter, use the values for B1 and B2 plotted **above** A1 on the original graph. This is indicated in Figure 5-16 with dashed arrows.

There are numerous possible outcomes in a complex experiment—slope of A (+, -, or 0), slope of B (+, -, or 0), and the interaction (yes or no). As an exercise, try your hand at reading the graphs in Figure 5-17 and replotting the data in the blank graphs to confirm your reading of the original graphs.

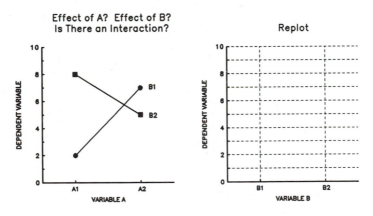

Figure 5-17 *(continues)*

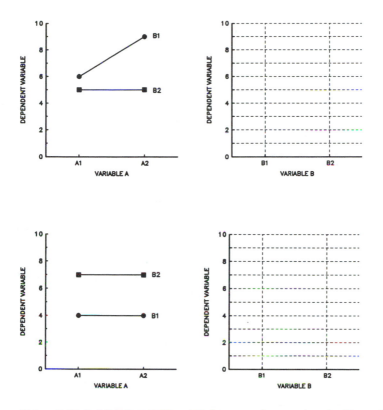

Figure 5-17 Practice in replotting data from complex experiments. Try reading the original graph for main effects of variables A and B and for any interaction, and then see if the replot confirms your initial reading.

F. GRAPH PROGRAMS FOR THE COMPUTER

The task of creating graphs for research papers and other publications has become computerized. There are a number of commercial graphics programs available for creating line graphs and other types of graphs. We briefly describe what graphics software does. We then give a quick overview of two popular graphics programs, Cricket Graph (which runs on Apple Macintosh computers) and Axum (which uses IBM PC compatible computers).

Computer graph-making programs are usually designed to perform these general functions: (1) creating and editing data tables, (2) creating and modifying graphs of the data, (3) storing data tables or graphs as files on a floppy diskette, hard disk drive, or other storage medium, and

(4) displaying graphs on a computer video display and printing or plotting graphs on paper. There are usually a number of optional ways to perform any of these basic functions.

To create a data table, we enter values of X and Y variables from the keyboard into a set of rows and columns appearing on the computer's video terminal. Most programs allow you to enter labels for the variables as well as numbers. Typically, each column of the data table represents a different variable, while the rows of the table are the values of the variable to be plotted. In creating a table of data for a line graph, one can have the first column of data (or labels) represent the X variable, while the second column represents values of the Y variable. Adding additional columns of data allows you to plot several Y variable relationships against the X variable (e.g., the graph can include several data curves showing the effects of more than one independent variable).

Once a data table has been entered, it may be edited from the keyboard at will, deleting values, or adding or rearranging rows and columns of data. Many graph programs allow you to transform the data in a variety of ways. For example, with a single typed command, one might take the square root of each value of a variable in a column, or multiply each value by 10, and so on. Some of the possible transformations are quite sophisticated. A program may even allow you to transform a variable by plugging it into a mathematical equation that you specify.

At the data entry and editing stage, it may also be possible to import data from another computer program such as a spreadsheet, a program used to manipulate rows and columns of data. Further, the program may allow you to export your data table for use by another program.

At the next stage, a data table you have created is turned into a graph. The graph creation and editing component of a graphics program probably offers the user the longest list of options. First, you will specify to the program what type of graph you wish to create. The list of choices is impressive: line graphs, bar graphs, scatterplots, pie charts (where segments of a circle represent proportions or percentages of a whole), and three-dimensional graphs (where a third variable is plotted on a third axis of a graph that is drawn in perspective) are typical choices.

Suppose you have simply elected to make a line graph from a table of X-Y data. In many graphics programs, you can specify an amazing number of aspects of the graph, including the relative length of the axes, the plotting symbols for data points, whether or not a data box is drawn, the type face and size of axis labels, the numbering of the axes and the spacing of tick marks, the content and location of labels, data legends and other lines and markers, and the style of lines (dashed, solid, and so on) connecting data points.

At any point in the process of entering data or creating graphs, most programs allow you to save your work in the form of a data file or graphics file that is written on a floppy disk or a hard disk connected to the computer. This is a wonderful feature, since it allows you to retrieve and edit files at will. Revisions and updates of graphs are possible and one can work on a graph until you are sure that it's right.

Once you have created a graph, you can obtain a hard copy of the graph from a printer. The highest-quality graphs come from laser printers with extremely high print resolution. But more conventional dot-matrix printers can also produce good quality copies of graphs. Most programs also support the use of plotters in which ink pens are moved mechanically across the surface of the paper. Most graph programs will give you choices about the final size, orientation, and position of your graph on the printed page.

The key pleasure of using a computer program to make a graph can be summed up with one word: **options**. At each stage of the process, most programs offer you a long menu of options allowing you to customize your graph in any way you please. At the same time, most programs incorporate a set of default parameters, specifications for the graph that will be used if you choose not to choose. A default line graph, for example, would have certain standard axis dimensions, plotting symbols, and so on, decided on by the programmers who created the software program.

I would like to conclude this section by mentioning the general characteristics of two popular computer graph programs that I have personal experience with, Cricket Graph and Axum.

Cricket Graph (Cricket Software, 1985) is a program written for the Apple Computer, Inc. line of Macintosh computers. Like many other Macintosh programs, it is **mouse driven**, meaning that instead of typing keyboard commands, the observer "points" to program options on the video screen and "clicks" on them using the moveable, table-top input device called a mouse. In Cricket Graph, the mouse can be used to "click" on actual components of a graph that appears on the screen (such as an axis scale or a data point) to modify that component directly, without resorting to more general menus or lists of options displayed on the screen. Cricket Graph also allows you to stretch or shrink the proportions of a graph by "dragging" it around on screen with the video-displayed mouse pointer. Cricket Graph thus offers the opportunity for hands-on interaction with a graph as it is displayed on the screen in front of the user. The program also incorporates a variety of functions in **menus** that allow the user to perform such advanced functions as data table transformations (including sorting and ordering data) and curve and line fitting of data points on graphs. Cricket Graph also gives the user great flexibility in changing the size, position, and so on of the printed graph, especially when the hard copy device is a laser printer.

Axum (Trimetrix, Inc., 1989) is a program written for IBM-compatible computers (PC, XT, AT, PS/2, and so on). It is a menu-driven program, in that options are selected from menus that pop up as overlays on the video screen. The strength of Axum rests in the incredible number of options that the user has as to type of graph, graphic format, data analysis including statistical analyses, linear regression, and analysis of variance (ANOVA), and print options. Axum is typical of full-featured graphics programs that allow one to become truly creative in the area of graphic design.

This brief overview of computer graphics does not do justice to the wide variety of advanced (but simple to use) graphic-making programs that are now available. Any college or university computing center can offer you graphics software capable of performing most of the functions and options I have mentioned.

CONCLUDING REMARKS

Graph literacy is an important skill that gives access to knowledge of experimental findings, data relationships, and trends over time. An understanding of graphs leads naturally to increased comprehension in science, technology, and commerce. I hope that *Graph It!* has either launched you into graphicacy or has helped you to polish (and appreciate) the graph literacy you already have. If you want to forge on to true graphic sophistication, the reference list at the end of the book contains notations about more advanced texts on graphic design and theory. If you are happy with the graphicacy that you have, have fun using it!

References

American Psychological Association. (1983) *Publication Manual*, 3rd. ed. Washington.

Baddeley, A. (1990) *Human Memory: Theory and Practice*. Boston: Allyn and Bacon.

Balchin, W. G. V. and Coleman, A. M. (1966) Graphicacy should be the fourth ace in the pack. *The Cartographer* 3, 23–28.

Bowen, R. W. (1981) Latencies for chromatic and achromatic mechanisms. *Vision Research* 21, 1457–1466.

Cleveland, W. S. (1984) Graphs in scientific publications. *The American Statistician* 38, 261–269.

Cleveland, W. S. (1985) *The Elements of Graphing Data*. Monterey, CA: Wadsworth. Note: This is an excellent advanced text on graphic theory and practice, and includes a summary of the author's work on graphical perception. The book also discusses a number of advanced graphic forms that supplement the line graph.

Cole, S. and Fiorentine, R. (1988) The formation of public opinion on complex issues: The case of nuclear power. In H. O'Gorman (ed.), *Surveying Social Life*. Middletown, CT: Wesleyan University Press.

Cricket Software, Inc. (1985) *Cricket Graph: Presentation Graphics for Science and Business*. (Software manual; software by J. Rafferty and R. Norling). Philadelphia: Cricket Software, Inc.

Ebbinghaus, H. (1964) *Memory: A Contribution to Experimental Psychology*. New York: Dover. (Originally published 1885)

Freeman, S., Walker, M. R., Borden, R., and Latané, B. (1975) Diffusion of responsibility and restaurant tipping: Cheaper by the bunch. *Personality and Social Psychology Bulletin* 1, 584–587.

Latané, B. (1981) The psychology of social impact. *American Psychologist* 36, 343–356.

National Center for Health Statistics (U.S. Department of Health and Human services). (1990) Annual summary of births, marriages, divorces, and deaths United States 1989. *Monthly Vital Statistics Report* 38, No. 13.

Playfair, W. S. (1801) *The Commercial and Political Atlas*, 3rd ed. London: J. Wallis.

Porter, H. (1939) Studies in the psychology of stuttering: XIV. Stuttering phenomena in relation to size and personnel of audience. *Journal of Speech Disorders* 4, 323–333.

Schmid, C. F. (1983) *Statistical Graphics: Design Principles and Practices*. New York: John Wiley & Sons. Note: This book and the text by Schmid and Schmid (1979) are excellent sources on the design of a variety of graphic forms. The books are especially rich in examples of time series graphs.

Schmid, C. F. and Schmid, S. E. (1979) *Handbook of Graphic Presentation*, 2nd ed. New York: John Wiley & Sons.

Stevens, S. S. (1962) The surprising simplicity of sensory metrics. *American Psychologist* 17, 29–39.

The World Almanac and Book of Facts 1991. M. S. Hoffman (ed.). New York: Pharos Books (Scripps Howard Co.).

Trimetrix, Inc. (1989) *Axum: Technical Graphics and Data Analysis*. (Software manual by C. Chuka; software by L. E. Edlefsen, D. O'Donnell, S. Ranney & F. Weinberg). Seattle: Trimetrix, Inc.

Tufte, E. R. (1983) *The Visual Display of Quantitative Information*. Cheshire, CT.: The Graphics Press. Note: Tufte's sophisticated treatment of graphic theory and practice is a delight both to read and to look at. The author traces a history of graphic application extending back to William Playfair (1801) using a number of interesting and amusing visual examples. The book assumes some degree of graph literacy, but it can truly be savored after you have *Graph It!* under your belt.

Index